PARALLEL SUPERCOMPUTING: METHODS, ALGORITHMS and APPLICATIONS

⊞WILEY SERIES IN PARALLEL COMPUTING

SERIES EDITORS:

J.W. de Bakker, *Centrum voor Wiskunde en Informatica, The Netherlands*
M. Hennessy, *University of Sussex, UK*
D. Simpson, *Brighton Polytechnic, UK*

Carey: Parallel Supercomputing: Methods, Algorithms and Applications

de Bakker (ed.): Languages for Parallel Architectures: Design, Semantics, Implementation Models

Axford: Concurrent Programming: Fundamental Techniques for Real Time and Parallel Software Design

Gelenbe: Multiprocessor Performance

Treleaven (ed.): Parallel Computers: Object-oriented, Functional and Logic

PARALLEL SUPERCOMPUTING: METHODS, ALGORITHMS and APPLICATIONS

EDITED BY

Graham F. Carey

University of Texas at Austin

JOHN WILEY & SONS

Chichester · New York · Brisbane · Toronto · Singapore

Other Wiley Editorial Offices

John Wiley & Sons, Inc., 605 Third Avenue,
New York, NY 10158-0012, USA

Jacaranda Wiley Ltd, G.P.O. Box 859, Brisbane,
Queensland 4001, Australia

John Wiley & Sons (Canada) Ltd, 22 Worcester Road,
Rexdale, Ontario M9W 1L1, Canada

John Wiley & Sons (SEA) Pte Ltd, 37 Jalan Pemimpin   05-04,
Block B, Union Industrial Building, Singapore 2057

British Library Cataloguing in Publication Data available

ISBN 0 471 92436 9

Printed and bound in Great Britain by Courier International, Tiptree,
Essex

Preface

Parallel computing has rapidly evolved from concept to a practical reality in only a few years. The recent advances in microelectronics and particularly the capability to produce inexpensive high speed processors, has led to an unusual situation: in some respects the study of parallel methods and algorithms lags the hardware development.

This has brought to the fore the need for more sophisticated compilers to automatically parallelize existing software. Further, we must develop new parallel numerical algorithm library routines. Finally, the whole issue of what is or is not efficient comes into question when different architectures (ranging from a few high-performance processors to many simpler processors) are considered on shared or distributed memory systems.

These pressing needs related to methods and algorithms for parallel and parallel-vector scientific computing have spurred research in this area and motivated the present monograph. The book has been organized to provide first background material and then lead into several inter-related fundamental topics. Chapters were written by invited contributors with this goal in mind. The authors also first exchanged ideas at an intensive 2-day workshop on the subject that was held in October 1988 in conjunction with the Numerical Analysis Conference in Austin honoring David Young. Through the workshop interaction and careful attention to the coordination of chapter materials with contributors, a well integrated approach to the subject has arisen.

In the first two chapters parallel speedup, Amdahl's law and some details concerning computer architecture configurations are introduced. Parallel scientific library needs, machine independent algorithms and software issues are then taken up. Next, we consider specific methodology designed to exploit "divide and conquer" domain decomposition strategies and element-by-element parallel-vector techniques. Methods for solution of boundary and evolution problems described by partial differential equations usually lead to large sparse systems of algebraic equations. In fact, much of the computational work associated with approximate solution stems from linear system solution. Accordingly, elimination and iterative solution schemes and software on parallel processors are then investigated. In the final chapters the methodology and algorithms are brought to bear on some representative applications to large scale complex problems from computational mechanics, viscous fluid flow, and reservoir simulation.

The methods and algorithms discussed here are basic to large scale scientific computing. The applications are quite representative and of general interest. Finally, the issues related to parallel performance on present and future systems are fundamental. Hence, this volume should not only be of interest to researchers active in these areas, but also to students and others desiring to become familiar with the subject. The organization of the aterial is such that it should also be of pedagogical value in a graduate course or seminar.

I would like to express my appreciation first of all to the contributing authors. Secondly, the U.S. Department of Energy has supported related research by Drs. Young, Kincaid, Sepehrnoori, myself and our graduate students. Finally, I would like to thank Belinda Treviño for the care she exercised in typing and formatting the manuscript.

Contents

Chapter 1

Performance Limits for Parallel Processors

Peter C. Patton[*]

1.1 Introduction

A number of computational science and engineering requirements have been expressed in recent years leading to strong current interest in the development of a "teraflop" computer. Such a machine would be able to deliver a sustained performance of 10^{12} floating point operations per second (flops). While today's leading super-performance computers offer peak performance in the one to ten gigaflop (Gflop) range, they actually deliver about 100 megaflops (Mflops) on most application codes. If 10^{12} flops is needed in the next decade and today's top super computers deliver 10^8 flops, how may one expect the factor of 10,000 performance deficiency to be resolved?

It appears that, short of some unexpected breakthrough in radical new technologies like superconductors or optics, semiconductor technology is reaching a point of diminishing returns. It took longer than expected to break through the 10ns barrier and still only a few manufacturers are delivering systems with clock times in the 4 to 9 ns range. Improvements in GaAs technology may allow the Cray-3 to operate at 2ns, and NEC is expected to reduce the clock on the forthcoming SX-3 to 3ns. The ETA-10G is available with

[*] NASA/JPL, Pasadena, CA.

liquid nitrogen cooled CMOS at 7 ns. Both Cray and Chen have suggested machines from their respective hands running at 1ns in the early 1990s.

Vector processing began with the Control Data STAR-100 in the early 1970s but first achieved commercial success with the Cray-1 in 1976. It took nearly ten years for vector processing to "catch-on" in scientific computing. It was advanced as an architectural technique by the Fujitsu VP- 200/Amdahl VP-1200 and the ETA-10 but reached its zenith in the NEC SX-2. The latter machine has four vector pipes overlapped with two scalar pipes invoked automatically without programmer intervention. While there may still be a way to go in perfecting FORTRAN compilers to automatically optimize codes for execution on vector processors, this architectural technique seems also to be reaching a point of diminishing returns.

Parallel processing is held forth as the great hope for a performance break-through, and after two decades of research and some significant development effort, is now available in the form of commercial hardware. In fact, many computers today have more parallelism than one knows how to apply to a single problem. As always, system and application software lag hardware development, but the widespread availability of parallel processors at every stage of technical computing including workstations, superminicomputers, minisu-percomputers, mainframes and supercomputers themselves has accelerated software development in both the research laboratory and the commercial marketplace. The question addressed here is whether there are limits on the performance gain that may be expected from parallel processing independent of the complexity of writing effective concurrent programs to execute concurrent algorithms.

1.2   Parallel Speedup

Figure 1.1 illustrates some speedup characteristics conjectured for parallel processing. The straight line at 45° represent linear speedup, i.e., P processor performance on a single task for P processors. While this is an ideal, there exist even a few algorithms showing superlinear speedup. Unfortunately this is rarely the case. While numerical methods before the advent of the electronic digital computer were highly concurrent, designed to be executed by a team of analysts working together on a large tableau with many small mechanical calculators, they were later "sequentialized" for use on early digital computers.

It has been widely noted that sequential codes do not run well on parallel computers. The $Log_2 P$ characteristic in Figure 1.1 represents Minsky's Conjecture, dating from the early days of research on Single Instruction Multiple Data (SIMD) nearest-neighbor multiple processor systems (Minsky and

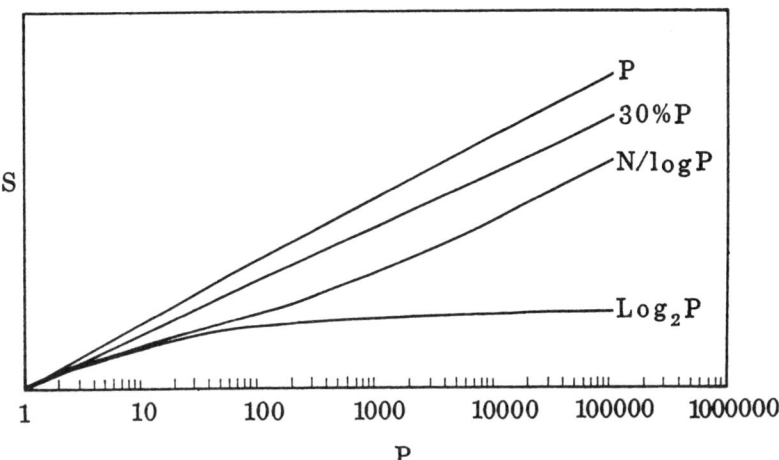

Figure 1.1: Parallel Processor Speedup: System Performance (S) versus Number of Processors (P).

Papert [7]). Here it was argued that some of the most favorable concurrent algorithms, such as binary search, branch and bound, etc. would use all of the processors on the first iteration, half on the second, a fourth on the third, etc.; hence, binary logarithmic performance overall. This conjecture seems much too pessimistic in light of today's experience. However, Minsky observes that for most heuristic based AI applications, what the computer does next usually depends on what it has just done, therefore parallel processing may not provide dramatic performance improvements.

The $P/\log P$ characteristic in Figure 1.1 represents the performance forecast from Amdahl's "Law": if a computer has two speeds or modes of operation during a given calculation, the slow mode will limit overall performance even if the fast mode is infinitely fast (Amdahl [1]). Interpreting Amdahl's Law as a simple rate problem makes it obvious. That is, the system rate R is given by $(1/R = 1/R_1 + 1/R_2)$ where R_1, R_2 are the subsystems slow and fast rates. That is, the system performs at the harmonic average of the subsystem rates. Figure 1.2 illustrates the Law from another perspective. This example is attributed to Ware, about 1964, in the literature (Patton [9]). It illustrates the sensitivity of parallel processor performance to the amount of a code that is intrinsically sequential. For example, in a 64 processor system, if 10% of the code is sequential, one may expect only 12 processor performance, but if only 2% is sequential, one may expect 40 processor speedup from 64 processors.

Recently Amdahl had the opportunity to reinterpret his law from the perspective of current issues in parallel computing (Amdahl [2]) Figure 1.3 is based on the chart he presented in his keynote speech to the Second Inter-

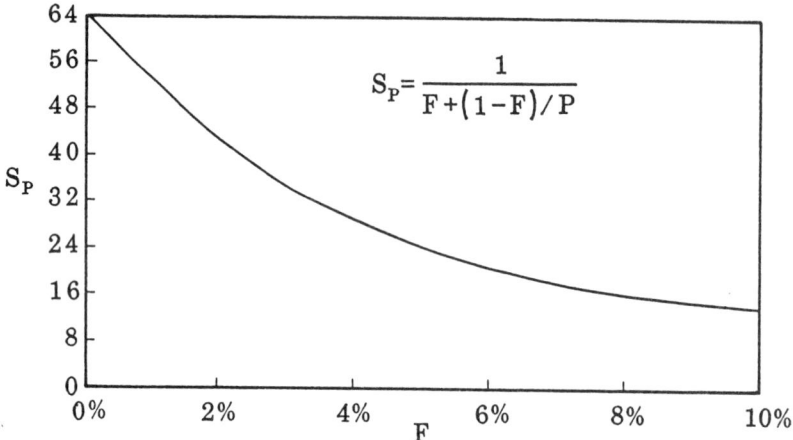

Figure 1.2: Example of Amdahl's Law: Speedup (Sp) over uniprocessor versus serial fraction (F) of computation for system with $P = 64$ processors.

national Supercomputing conference in Santa Clara, California in May 1987. He noted that the great majority of programs in technical computing exhibit 50-70% concurrency available for exploitation by either a vector or parallel architecture. Thus, even if one had 1000 or more processors, the performance characteristic would show a saturation knee at about 150 or 160 processors. He did note that some problems exhibiting an intrinsic physical or logical decomposability showed 85-94% speedup, but were exceptional.

1.3   The Sandia Experiments

A parallel processing research program at Sandia Laboratories was described in the New York Times on March 15, 1988. The Times noted the widely held belief that it was impossible to get more than 100 times the speedup of a given single processor by parallel processing, employing any number of processors. Also, the Karp prize had been established for the first research team to gain a factor of 200 or better. The Times reported that Gustafson, Montry, and Benner of Sandia had done three experiments, all with performance results better than 1,000 times uni-processor performance, and had won the Karp Prize. The Sandia researchers observed that faster computers have usually been employed to do new, large problems rather than just to solve old smaller problems faster. Hence, it was argued that the problem size should be scaled-up as the number of processors is increased. A series of numerical experiments was carried out to solve scaled-up problems

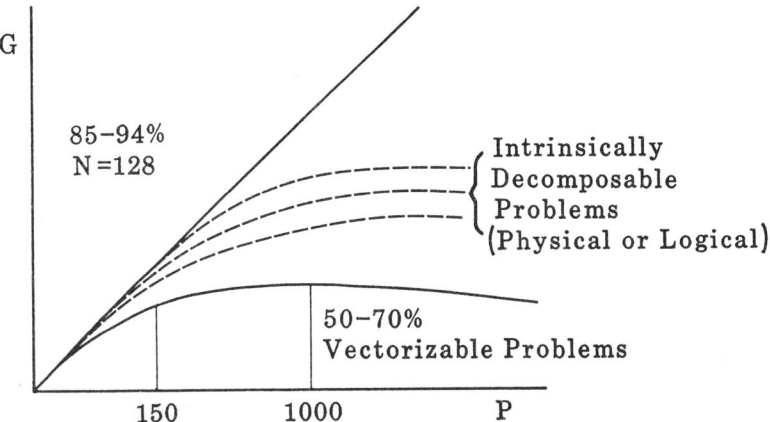

Figure 1.3: Amdahl's Law Revisited: Global performance (G) on a specific problem versus number of processors (P).

on scaled-up computers to support this reasoning ([3,4]) as described in a later chapter.

The parallel processing research focus at Sandia is on a 1,024-processor hypercube environment. The large number of processors forces examination of every sequential aspect of a program. It also leads to re-examination of the traditional paradigm for measuring parallel processor performance. It was demonstrated that it can be easier to achieve a high degree of parallelism than might at first be inferred from Amdahl's Law. It is often stated that production scientific programs have a substantial inherent serial component, s (several percent), that limits the usefulness of the parallel approach to an asymptotic speedup of $1/s$. The Sandia results indicate that this is not necessarily the case. They show that s can be made quite small on practical problems by a variety of techniques that reduce nonoverlapped communication, load imbalance, message startup time, sequential operation communication, and sequential operation dependency. Also, when problem size is scaled in proportion to the number of processors, s effectively decreases, reducing the barrier to speedup as the number of processors is increased.

At Sandia, massively parallel algorithms that achieve very high parallel efficiencies (98.7 percent or better) were applied to three large-scale scientific applications. The applications were drawn from wave mechanics (using explicit finite differences), fluid dynamics of an unstable flow (using the flux-corrected transport technique), and beam strain analysis using finite elements and conjugate gradient iteration. In these calculations, when the problem size is fixed, the serial component, s, ranges from 0.0006 to 0.001, resulting in parallel speedups of 502 to 637. When the problem size is increasesd with the number of processors, the equivalent serial component drops to between 3 and 10 parts per million, resulting in parallel speedups of 1,009 to 1,020 on a 1,024-processor NCube. This reinterpretation of scaled speed-up with increasing problem size is the essence of the argument.

In reducing the serial component to such a small number, several previously masked sources of parallel efficiency loss became apparent. New effects of massive parallelism were observed that were unimportant on systems of fewer than 1,000 processors. These effects include inefficiency caused by redundant operations, spurious load imbalance induced by hardware defects, and data-dependent Mflop rates. Future massively parallel hardware designs may find these effects increasingly important as the number of processors in ensembles is increased, and the traditional problems of parallel efficiency loss are solved.

If P is the number of processors, s is the amount of time spent (by a serial processor) on serial parts of the program, and p is the amount of time spent (by a serial processor) on parts of the program that can be done completely in parallel, then Amdahl's Law gives:

$$\text{Speedup} = \frac{(s+p)}{\left(s+\frac{p}{P}\right)} = \frac{1}{\left(s+\frac{p}{P}\right)} \tag{1.1}$$

where total time is normalized to $(s+p)$ equals one. For $P = 1,024$ this is a steep function of s near $s = 0$.

The expression in (1.1) is based on the implicit assumption that p is independent of P. However, one does not generally take a fixed-sized problem and run it on various numbers of processors; in practice, a scientific computing problem scales with the available processing power. The main constraint is not the problem size but rather the amount of time a user is willing to wait for an answer. When given more computing power, the user expands the problem to use the available hardware and reduce problem flow-time.

As a first approximation, the Sandia results indicate that the parallel part of a program scales with the problem size. Times for program loading, serial bottlenecks, and I/O that make up the s component of the application do not scale with problem size. The amount of work that can be done in parallel varies linearly with the number of processors. In fact, the inverse of Amdahl's paradigm was introduced: rather than ask how fast a given serial program would run on a parallel processor, they asked how long a given parallel program would have taken to run on a serial processor. Using s'' and p' to represent serial and parallel time spent on the parallel system with $s' + p' = 1$, then a single processor requires time $s' + p'P$ to perform the task.

This reasoning gives an alternative to Amdahl's Law:

$$\text{Scaled speedup} = \frac{(s' + p'P)}{(s' + p')} = P + (1 - P)s' \tag{1.2}$$

In contrast, this function is simply a line of moderate slope, $1 - P$. When speedup is measured by scaling the problem size, the scalar fraction, s tends to shrink as more processors are used. It is thus easier to achieve efficient parallel performance than is implied by Amdahl's law, and performance as a function of P is not necessarily bounded in the manner previously inferred.

Measurements of parallel performance may best be displayed as a function of both problem size and ensemble size. Two subsets of this domain have received attention in the parallel processing community. The first subset is called the fixed-sized speedup line in Figure 1.4. Along this line, the problem size is held fixed, and the number of processors is varied. On shared-memory machines, especially those with only a few processors, this is reasonable, since all processors can access memory through a network transparent to the programmer. On local memory machines, fixing the problem size creates a severe constraint, since, for a large ensemble, it means that a problem must run efficiently even when its variables occupy only a small fraction of available memory.

For ensemble computers, the scaled speedup line in Figure 1.4 is an alternative computational paradigm. It is the line along which problem size increases with the number of processors. The computation-to-communication ratio is higher for scaled problems. Their performance is compared with a hypothetical processor node that has direct single processor access to all of the real random-access memory of the machine. This hypothetical processor performance is numerically equivalent to the ratio of measured Mflop rates.

1.3.1 Moler's Law

Moler [8] when at Intel, developed an extension of Amdahl's Law for multiprocessors that illustrates Sullivan's main contention, that parallelism can attain desired speedup for sufficiently large computations, and also anticipated the Sandia results. Moler defines α as the Amdahl fraction—that part of an algorithm that is not parallelizable. Then the computation time T_P and speed-up S_P with P processors become

$$T_P = \frac{1 - \alpha T_1}{P} + \alpha T_1 \tag{1.3}$$

$$S_P = \frac{P}{1 + (P - 1)\alpha} \le \frac{1}{\alpha} \text{ for all } P > 1 \ . \tag{1.4}$$

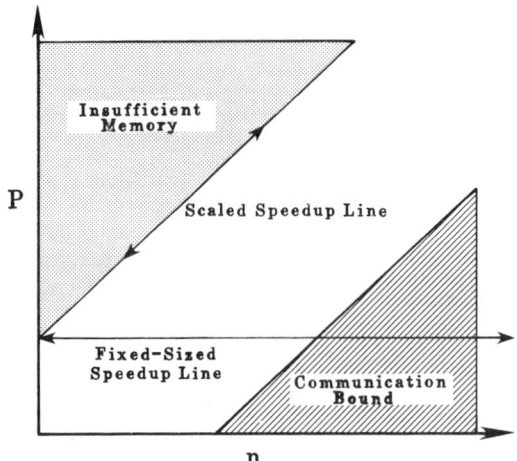

Figure 1.4: Ensemble Computing Performance Pattern: Log-log graph of problem size (n) versus number of processors (P).

For example, if 5 percent of an algorithm is not parallelizable, then Amdahl's Law implies that the maximum speedup is 20, no matter how many processors are used.

If one considers not a given problem and more processors but rather a sequence of problems of increasing size, n, the Amdahl fraction α, can be shown to be dependent on problem size. Moler defines an effective parallel algorithm as one for which $\alpha(n) \to 0$ as $n \to \infty$. If P is fixed in the expressions for speedup above and $n \to \infty$, then $S_P \to P$.

For example, an effective algorithm for matrix-vector multiplication $y = Ax$ distributes a large matrix, A, by rows A^i across multiple processors and stores the vector, x, in one processor. Then row dot products with x are processed in parallel.

Algorithm 1.1 Parallel Matrix-Vector Product.

$$\begin{array}{lll}
\text{To Compute} & y = Ax & (1.5) \\
\text{Algorithm} & \text{Send } A \text{ to all processors} & \\
& \text{Send } x \text{ to all other processors} & \\
& \text{All processors simultaneously:} & \\
& \quad \text{for } i \text{ in my rows} & \\
& \quad y^i = A^i x & \\
& \text{Accumulate } y \text{ in destination processor} &
\end{array}$$

Moler's analysis of the algorithm is as follows: For P processors, matrix size n and $m = \lceil n/P \rceil$ the maximum number of rows in any processor, the following contributions to the Amdahl fraction are obtained: (1) parallel overhead in distributing x; (2) load imbalance if n is not divisible by P; (3) parallel overhead in accumulating y.

The useful work may be totaled as:

$$\text{send } x \sim n \log P \tag{1.6}$$
$$m \text{ dot products} \sim mn$$
$$\text{accumulate } y \sim n \log P$$
$$\text{Total} \sim O(mn) = O(n \log P)$$
$$\sim O\left(\frac{n}{P}\right)^2 + O(n) + O(n \log P).$$

Thus

$$\alpha(n) = \frac{O(1 + \log P)}{n} \rightarrow 0 \qquad \text{as } n \rightarrow \infty \tag{1.7}$$

and the distributed matrix-vector multiplication algorithm is effective asymptotically in the sense described.

1.4 Sullivan's Theorems

Sullivan [10,11] proves two propositions: first, that the maximum speedup achievable from a parallel computer with P processors executing a computation of concurrency C is Minimum $[P,C]$; second, that for every P, there exists a realizable parallel computer that provides a speedup $O(P)$ for every computation that has a concurrency $O(P \log P)$.

The parallel processing goal is to improve the execution speed for a single large problem. Ideally, this speedup would be applicable to all very large problems and would therefore have the same effect as an increase in the rate at which the computer can perform hardware operations, that is, the same effect as increasing the clock rate. An important goal is the achievement of an execution rate of one tera-op, or 10^{12} operations per second, by the early 1990s. If this were to be achieved on a conventional computer, the clock period would have to be less than one picosecond. Such an improvement in technology does not seem likely, so designers must rely on architectural techniques, which will have to provide a speedup factor of between 10^3 and 10^5. This will require a corresponding number of computing elements or processors. The basic questions are how many processors can be organized and coordinated to perform a single large computation, and how efficient can this structure be made?

Sullivan considers the construction of a parallel machine that exploits all forms of concurrency. It is widely thought that all very large computations exhibit abundant concurrency, and, if this is the case, the desired speedup can be attained for sufficiently large computations with moderate clock rates and highly parallel structures. Parallel structures which speed up computing by exploiting concurrency have been employed for two decades. They include pipelines, multiple function units, wide-word structures, vector units, array processors, and multiprocessors. Each is restricted to some particular form of concurrency and is therefore applicable only to certain aspects or parts of typical computations. Therefore, these techniques have not provided the solution to the general problem considered here. To achieve the desired speedup, it is necessary to exploit effectively all forms of concurrency. Kuck [5] was the first to study concurrency independent of hardware structure. He used the tool called a dependency graph, or execution graph, to depict the concurrency available.

A concurrent computation is called well-defined when the operations it performs, together with their partial ordering, is given. The partial ordering is defined by the condition that each operation requires operands, either from input data or from results generated from previous operations. A well-defined computation, therefore, generates an execution graph in which nodes represent operations and arcs represent operands. Each operator together with its operands is an "atomic task". The execution graph may be put into standard form by displaying atomic tasks in rows, so that each atomic task in a row depends on some atomic task in the previous row. Figure 1.5 illustrates the execution graph for a concurrent program.

The size S of the execution graph is the total number of operations performed during the execution. The depth D is the number of nodes in the longest path through the execution graph, or the number of rows in a graph exhibited in standard form. Concurrency is defined as the ratio S/D. The number of atomic tasks in C_i the ith row, is the instantaneous concurrency. There are C_i opportunities for parallel execution. In fact, Sullivan shows that the average number of opportunities for parallel execution is equal to the concurrency.

The execution graph is a static representation of the atomic tasks executed during computation. One can relate this graph to the speedup available from an ideal machine, which has P processors available to perform the computation represented by the execution graph. At each step in the computation, the ideal machine performs three functions: first, it identifies all the atomic tasks whose operands have already been evaluated (i.e., executable atomic tasks); second, it assigns processors to perform the operations; third, it sends

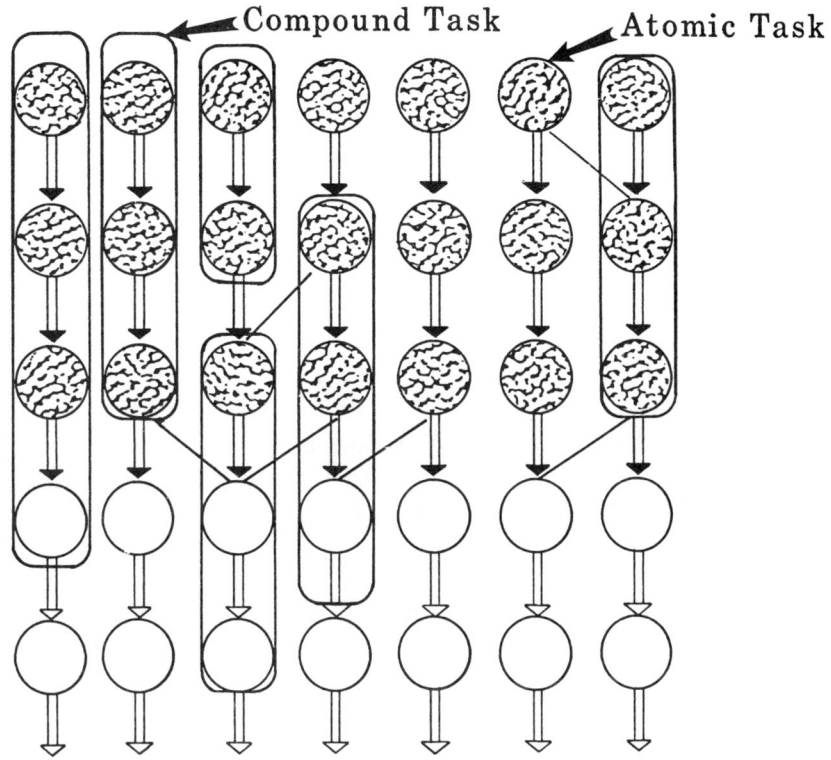

Figure 1.5: Execution Graph for a Concurrent Program.

the operands to the processors. Each processor then executes its atomic task, ideally in one unit of time. If the machine has enough processors, it assigns a processor to each task on each row of the graph, and each row is therefore completed in one unit of time. The entire computation is therefore completed in D time units, where D is the depth of the graph.

There is no universally accepted definition of a concurrent program, but Sullivan adopts a definition that is, in a sense, minimal. In this definition, the object code specifies all the operations to be executed in the computation and the order in which these operation are to be performed. (In the sequential case, this is the standard definition of a program). In the concurrent case, the ordering is partial, i.e., at any instant it specifies that a plurality of operations can be performed. Thus, at each instant during the computation, the stored program specifies the operations to be performed. This implies one and only one execution graph, once the input data is defined. The speedup of any parallel machine with P processors for some given execution is defined as the ratio of the time required with P processors. It is clear that the speedup can

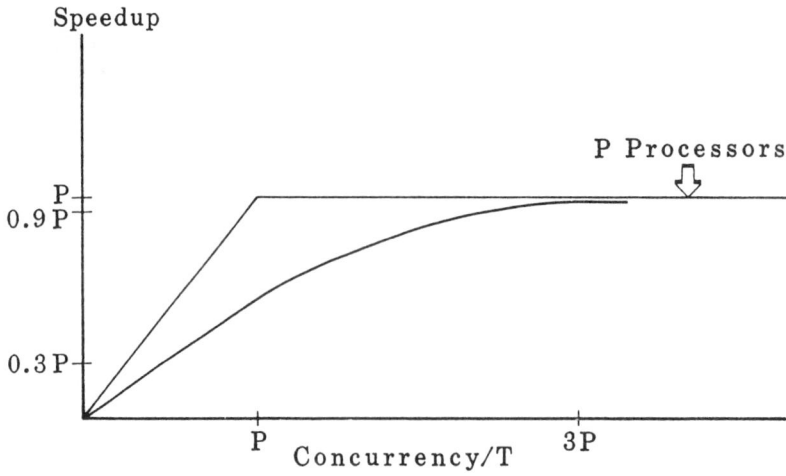

Figure 1.6: Performance Under Random Task Assignment.

never exceed P. If the ideal machine is limited to one processor, the time to perform the computation is S units of time, because there are S atomic tasks, and each is performed by a processor in one unit of time. Thus, the maximum speedup achievable from a parallel computer with P processors executing a computation of concurrency C is Minimum $[P, C]$.

Sullivan's argument for a lower performance bound employs fixed assignments of compound tasks to processors using a random procedure and the performance result is sketched in Figure 1.6. Let n_i be the largest number of executable tasks found in any of the P processors on the execution of the ith row of the execution graph. Then n_i is the time required to complete the ith row, and the total time t is

$$t = \sum_{i=1}^{D} n_i \qquad (1.8)$$

Also, the expected value, $\bar{t}$ is

$$\bar{t} = \sum_{i=1}^{D} n_i \qquad (1.9)$$

Among all graphs of concurrency C, the maximum value of t is attained for those satisfying

$$C_i = C \text{ for all } i, \qquad (1.10)$$

Thus

$$\bar{t} = D\bar{n} \qquad (1.11)$$

The statistic $\bar{n}$ is the expected value of the largest number of tasks in any of the P processors when C tasks have been assigned at random. The expected speedup is

$$\sigma = \frac{s}{t} = \frac{DC}{D\bar{n}} = \frac{C}{\bar{n}} \tag{1.12}$$

The parameter

$$\rho = \frac{\sigma}{P} \tag{1.13}$$

is introduced as a measure of processor utilization. Sullivan argues that, for any value of $r \le C$

$$\bar{n} < r - 1 + P \frac{e^{-\frac{C}{P}}}{\sqrt{2\pi \frac{C}{P}}} \left(\frac{eC}{rP}\right)^r \frac{1}{1 - \frac{C}{rP}} \tag{1.14}$$

If one chooses $r = \frac{C}{P\rho}$, and considers values for C and p such that the second term is less than 1, then

$$\bar{n} = \frac{C}{P\rho} \tag{1.15}$$

and the speedup is

$$\sigma = P\rho . \tag{1.16}$$

Thus, the speedup σ of the processor that makes the fixed assignments of compound tasks at random is

$$\sigma = P\rho \tag{1.17}$$

where ρ satisfies

$$P \frac{e^{-C/P}}{\sqrt{2\pi \frac{C}{P}}} (e\rho)^{C/P} \frac{1}{(1 - \rho)^2} \le 1 . \tag{1.18}$$

He observes that if $C = P \log P$, and $\rho \le 1/e$, the inequality is satisfied, and, therefore: For every P there exists a parallel computer that provides a speedup $O(P)$ for every computation that has a concurrency $O(P \log P)$.

1.5   Worley's Limit

Worley [12] shows that the computational cost of calculating a numerical approximation to the solution of a partial differential equation is constrained by the amount of data it must use in order to calculate a sufficiently accurate approximation. Under simple assumptions about the data function and the partial differential equation these constraints limit how quickly the approximation can be calculated. As ϵ, the bound on the error in the approximation, decreases, he demonstrates a lower bound on the parallel complexity that increases at least linearly as a function of $\log_2\left(\frac{1}{\epsilon}\right)$. This bound is tight in the sense that there exists a family of explicit linear algorithms with parallel implementations whose parallel complexity grows no faster.

The number of processors used to calculate the approximation must increase as the error bound decreases if the lower bound on parallel complexity is to be achieved. This usually entails an increase in communication cost, but the communication cost does not change its asymptotic behaviour as long as the parallel algorithm remains computation-bound. This is not always possible for small error tolerances when the approximation is calculated on a multiprocessor whose network "radius" grows faster than logarithmically. In particular, the parallel cost of optimal algorithms on d-dimensional grid architectures is determined by the communication capabilities of the architecture rather than by minimum parallel complexity. For given processor speed, there exists a logarithmic bound on the radius of a multiprocessor that ensures that the communication cost associated with the radius will not influence the behavior of the parallel cost of optimal parallel algorithms.

Worley notes that the assumptions employed to prove these results are merely sufficient, and not necessary. He contends that they are reasonable assumptions for *a priori* estimates on the error for mth-order accurate approximations to the solution of many important classes of linear scalar partial differential equations. Worley argues that the assumptions and analysis for linear scalar PDEs are also extendable for other classes of problems. For example, little is changed for systems of linear PDEs. Many linear problems also have integral representations for the solution operator for each element of the solution vector, where the kernels satisfy the usual assumptions.

For nonlinear problems the solution operator may not have a simple representation but most numerical approximations to nonlinear problems can be developed as solutions to a sequence of linear problems. The results then carry over if these linear subproblems satisfy Worley's assumptions. Also, data functions may be discontinuous for some applications; however, the data is generally smooth on all but a finite number of bounded smooth lower dimensional surfaces. The analysis follows through as before as long as the solution is also smooth except on a finite number of bounded smooth lower dimensional surfaces for this type of data. Finally, the analysis depends only on the solution operator having the required integral formulation. Thus, the results are not restricted to linear partial differential equations.

It is interesting to compare Worley's analytical results (Worley [12,13] with Kumar's experimental measurements of parallelism in computation-intensive scientific applications (Kumar [6]). Recall that Worley's bound on parallel complexity is a linear function of $\log_2 \epsilon^{-1}$, where ϵ is an upper bound on the error. Hence, parallel processing architectural gain is somewhere in the range of 1,000 to 10,000 processors depending on the communication cost of the algorithm. Kumar's experiments also show instantaneous peaks in program concurrency in the 1,000 to 10,000 range.

1.6 Conclusion

Experience in computational chemistry over the past two decades has exhibited a performance gain of 10^7, half of which came from computers and half from algorithms. Similar results have been achieved in computational aerodynamics. Researchers at the National Radio Astronomy Observatory have vastly improved their algorithms over the past decade while making them 100 times faster. Returning to the initial inquiry, we now have a better perception of the directions that would lead to a tera-flop machine in the next decade. Summarizing the considerations presented here, one can anticipate perhaps another factor of 4 to 10 speedup due to advances in electronic technology. Worley's limit and Kumar's experiments place an upper bound on the architectural gain from parallel processing in the 1,000 to 10,000 processor range. If these advances can be exploited in combination, it appears that a teraflop computer is feasible before the end of this century.

References

[1] Amdahl, G., "Validity of a Single Processor Approach to Achieving Large Scale Computing Capabilities," AFIPS Conference Proceedings, 30, 483–485, 1967.

[2] Amdahl, G., Keynote Address, Second International Supercomputer Conference, Santa Clara, CA, May 1987.

[3] Gustafson, J., "Re-evaluating Amdahl's Law," *CACM*, 31, 5, 532–533, 1988.

[4] Gustafson, J., G. R. Montry and R. E. Benner, "Development of Parallel Methods for a 1024-Processor Hypercube," *SIAM J. Sci. Stat. Comput.*, 9, 4, 609–638, 1989.

[5] Kuck, D. J., *et al.*, "Measurements of Parallelism in Ordinary FORTRAN Program," *IEEE Computer*, 7, 1, 29–59, 1977.

[6] Kumar, M., "Measuring Parallelism in Computation — Intensive Scientific Engineering Applications," *IEEE CT*, 37, 9, 1088-98, 1988.

[7] Minsky, M., and S. Papert, "On Some Associative, Parallel and Analog Computations," in *Associative Information Techniques*, Elsevier, 1971.

[8] Moler, C., Personal Communication, 1988.

[9] Patton, P. C., "Multiprocessors: Architecture and Applications," *IEEE Computer*, 28, 6, 29–40, 1985.

[10] Sullivan, H.W., "How Much Can We Speed-up Supercomputers by Exploiting Concurrency?," in Supercomputing Special Issue of *Interdisciplinary Science Reviews*, in press, 1989.

[11] Sullivan, H.W., *The Future of High Performance Computing*, CHoPP Computer Corp., November 29, 1988.

[12] Worley P.H., *Information Requirements and the Implications for Parallel Computation*, Ph.D. Thesis, Stanford University, STAN-CS-88-1212, June 1988.

[13] Worley, P.H., *Limits on Parallelism in the Numerical Solution of Linear PDEs*, Oak Ridge National Laboratory Report ORNL/TM-10945, October 1988.

Chapter 2

Overview of Current Developments in Parallel Architectures

*Oliver A. McBryan**

2.1 Introduction

Supercomputers are the key to the simulation of a wide range of important physical problems. Such simulations typically require large numbers of degrees of freedom to provide sufficient resolution, particularly when engineering accuracy, rather than simple qualitative behavior, is required. In many cases one is currently limited by available computer resources, rather than by an understanding of the underlying physics.

A prime example of the needs for massive computer simulation is exhibited by the case of weather and climate forecasting. Global weather models represent the atmosphere as a three dimensional grid laid out along latitude, longitude and vertical directions on the sphere. The most detailed weather simulations performed to date use about 200 grid points in each horizontal coordinate direction and perhaps 20 in the vertical. Such resolution is woefully inadequate. To place this in context, note that with such a resolution the state of Colorado would be represented by a single point, completely ignoring the variations in mountain elevation within the state.

Even more challenging than the weather modeling problem is the prediction of flow in the ocean. The importance of oceanographic simulation is

---

*University of Colorado, Boulder, CO.

borne out by the wide-spread disruption caused by the return in 1986 of El Nino, a cyclical tropical current in the Pacific Ocean. El Nino is believed to have dominated world weather conditions for close to two years before it finally decayed. The primary difficulty with ocean modeling is caused by the enormous range of scales required to effectively simulate an ocean. While the solution undoubtedly lies in the direction of better turbulence models to describe the flow, the use of massive supercomputer simulations may well be essential to discovering such better models.

As another example, it is very desirable to simulate accurately the flow of air over a plane. Current aircraft design strategy involves extensive use of wind tunnel testing. However wind tunnel testing is limited with respect to aircraft size and Mach number, although extrapolations from smaller scale models can overcome some of the limitations. Planned wind tunnel testing for the Boeing 7J7 was greatly reduced by advances in computational aero-dynamics, substantially curtailing 7J7 development time and, consequently, costs. But the computational techniques now in use do not simulate the complete physics for the flow past the entire aircraft; they model various aspects of the flow that, when combined, give guidance to the design, but not necessarily accurate solutions. The major limitation is that as more vehicle details are included in the simulation, the numerical grids become larger, requiring more processing power and memory.

The same difficulty is seen in oil reservoir simulation, in combustion studies, and wherever quantitative computations in three dimensions are performed. A non-inclusive list of application areas that would benefit substantially from increased computer power follows: Weather Modeling, Global Change and the "Greenhouse Effect", Aerodynamic Design, VLSI Design, Automobile Design and non-destructive testing, High Temperature Superconductivity, Pharmaceutical Industry, Molecular Modeling and "Designer Chemicals", The Human Genome sequencing project, Computer Vision and Image Processing, Seismic Processing, Oil Reservoir Simulation, Stability of Large Structures, Quantum Chromo-Dynamics, Transport Phenomena, Linear and Non-linear Optimization, and Cryptography.

While there are no absolute standards for comparing computer performance, the concept of peak floating point performance is one widely used criterion. This performance measure is generally described by the number of floating point operations per second ("flops") that the computer can deliver. For this purpose additions and multiplications are usually treated equally (as well as subtraction or division), whereas operations such as copying floating point data, or initializing data are ignored. A rate of one gigaflop (abbreviated Gflop) represents a processing power of one billion floating point operations per second. We will also use the names megaflop (or Mflop) for a million operations per second, and teraflop (or Tflop) for a trillion operations per second. Several current supercomputers have passed the one Gflop barrier. Major advances in many of the listed application areas are expected as soon as computer power increases to about 100 Gflops, and especially if one can reach the Tflops range. This would correspond to an increase of close to an order of magnitude in resolution in each of the coordinate directions of typical fluid simulations, compared to current vector machine capabilities.

Conventional supercomputers with one or a few processors are limited by various factors, including the need to dissipate energy in a small volume, effects of the finite speed of light, and bottlenecks related to memory access. It is widely believed that parallel computers provide the only hope of reaching this range of computer power in the near future. Furthermore, in most applications the cost per megaflop is a relevant issue. Massively parallel computers provide economies of scale not available to conventional computers larger than a personal computer. Parallel computers may in addition be built from lower cost technologies, because the individual processors need not be particularly powerful.

Because of these factors, parallel computers have been widely studied in recent years. Substantial research has been accomplished related to these machines, including both theoretical advances, involving algorithm design, and computational experiments. Hardware advances have reached the point where the fastest available supercomputers are now highly parallel machines McBryan [7].

The one great disadvantage of a parallel computer, is that it is much harder to program than a serial machine. Each processor must be assigned a distinct component of the work to be performed, and substantial synchronization of the processors is then required in order to ensure that the results from individual processors are merged appropriately. The difficulties of programming parallel machines have spawned a whole range of new research areas for computer science and are a primary reason why this area has been so dynamic in recent years.

In Section 2.2 we give an overview of the architectural and software approaches used by several current parallel computers. Node design and communication features are briefly discussed. In Section 2.3 we review in more detail some of the specific parallel architectures that are currently available or are that under development. These range from a few processors to massively parallel systems. For further details on several of these architectures and for examples of applications such as partial differential equation solution on these machines, we refer to our papers [4,6-11], and in particular to the following chapters in the present volume.

2.2   Overview of Parallel Systems

2.2.1   Classification of Parallel Computers

Parallel computers may be broadly categorized in two types — SIMD or MIMD[8]. SIMD and MIMD are acronyms for Single Instruction stream — Multiple Data stream, and Multiple Instruction stream — Multiple Data stream, respectively. In SIMD computers, every processor executes the same instruction at every cycle, whereas in an MIMD machine, each processor executes instructions independently of the others. The vector unit of a CRAY computer is an example of SIMD parallelism — the same operation must be performed on all components of a vector. Most of the interesting new parallel computers are of MIMD type which greatly increases the range of computations in which parallelism may be effectively exploited using these machines. However, this occurs at the expense of programming ease — MIMD computers are much more difficult to program than SIMD machines. Many current designs incorporate both MIMD and SIMD aspects — typically each node of an MIMD system is itself a vector processor.

Another easy categorization is between machines with global or local memories. In local memory machines, communication between processors is entirely handled by a communication network, whereas in global memory machines a single high-speed memory is accessible to all processors. Beyond this, it becomes difficult to categorize parallel machines. There is an enormous variety in the current designs, particularly in the inter-connection networks. For a taxonomy of current designs, see the paper of Schwartz [12].

While many interesting parallel machines involve only a few processors, we will concentrate in this paper on those machines which have moderate to large numbers of processors. Important classes of machines such as the CRAY X-MP, CRAY-2 and ETA-10 are therefore omitted from the subsequent discussions.

2.2.2   A Partial List of Multi-processors

There are numerous parallel computer projects underway at this time, worldwide. While some of these projects are unlikely to lead to practical machines, a substantial number will probably lead to useful prototypes. In addition, several commercial parallel computers have been produced (e.g., ICL DAP, Denelcor HEP, Intel iPSC, NCUBE, FPS T-Series, Connection Machine, Symult, Meiko, Parsatec) and more are under development. One should also remember that the latest CRAY computers, (e.g. CRAY Y-MP and CRAY-2) involve multiple processors, and other vector computer manufacturers such as ETA Systems, NEC, Fujitsu and Hitachi have similar strategies.

Table 2.1 lists a selection of the parallel computers under development. This is just a sample of the projects mentioned above, but covers a wide range of different architectures. Beyond the simple classification into SIMD or MIMD computers we recognize a vast array of different approaches to the task of building a parallel architecture. We will now look at the reasons for this broad array by discussing some of the possibilities encountered for both node and communication facilities.

Table 2.1: Some Parallel Computer Projects.

AMT DAP-II	Caltech Hypercube
Intel iPSC hypercube	NCube Hypercube
Denelcor HEP-1	NYU/IBM Ultra-computer/RP3
Connection Machine CM-2	FPS T-Series
CRAY Y-MP and CRAY-2	ETA-10 Series
IBM 3090 Multiprocessor	Multiflow Trace
Goodyear MPP	MIT Data-flow Machines
BBN Butterfly	Wisconsin Database Machine
SUPRENUM-1 and 2	IBM GF-11 and TF1
Paralex Pegasus	Symult 2010
Myrias SPS-2	Cedar Project
Meiko Computing Surface	Parsetec
Evans & Sutherland ES-1	CMU iWarp
Flex	Alliant FX-8
Sequent Balance	Encore Multimax
CCI Navier-Stokes Machine	TERA

2.2.3 Node Design

Node design tends to be far less variable than other aspects of parallel computers. The main reason for this is that most architects have relied on off-the-shelf products for the node — standard microprocessors, floating point accelerators and memory chips. The advantage is that startup time for a project may be substantially reduced. Additionally there is usually a substantial body of low-level software available for such processors — software such as compilers, assemblers and debuggers. Thus we find that an enormous number of the current parallel computer products are based on one or more of the Intel 80386, Motorola 68020, Inmos T800 transputer and the Weitek floating point accelerators. Typically one of these microprocessors will be combined on a board with a floating point co-processor (e.g. 80387 or 68881), possibly a Weitek chip set and several megabytes of memory. Memory consumes substantial space, and current systems have in the range of 1 to 20 MBytes per node. Despite these general comments, it should be mentioned that some manufacturers have developed custom processors specifically for parallel computers. In the list above we would point to the DAP, NCube, HEP-1, CM-2, ES-1, iWarp and Navier-Stokes machines as examples. The iWarp is of particular interest here on account of the high level of integration used in the design of the custom processing element.

2.2.4   Communication Features

The range of inter-processor communication facilities is what really characterizes the differences in architecture among the various parallel machines. While we have previously distinguished the shared memory and distributed memory classes, one should observe that this distinction should not be taken too seriously. A distributed memory computer can certainly simulate a global shared memory and vice versa.

Communication pathways are typically built either from direct point-to-point connections, or from busses. Busses have the advantage that many processors may be serviced by one communication path, but have the disadvantage of slower bandwidth performance as the number of processors increases. With point-to-point connections, processors that are directly connected will have very efficient communication, but indirectly connected processors will likely incur substantial extra overheads including increased latency as well as lower bandwidth.

The most popular interconnection strategies involve simple symmetric arrangements including rings, meshes, hypercubes, trees and complete connections or crossbars. The prevalence of hypercube designs is explained by the fact that this architecture supplies substantial parallel bandwidth for many standard algorithms, for example the Fast Fourier Transform, while at the same time incurring only relatively modest fan-in and fan-out of connections which grow in number only logarithmically with the processors. Table 2.2 compares several simple topologies as a function of processor number P from the point of view of amount of wiring (difficulty of building), connectivity (ease of programming) and maximal path (efficiency of long-range communication). While cross bar switches are extremely difficult to build for large numbers of processors, they have tremendous flexibility in terms of efficiency and ease of use. It is conceivable that a technological breakthrough such as optical switching might allow cross bars to be built that would connect

Table 2.2: Properties of Interconnection Networks.

Network	Wires	Connectivity	Max Path
Cross Bar	P	P	1
1D Grid	P	2	P
2D Grid	2P	4	$2\backslash|P$
Binary Tree	P	1–3	2logP
Hypercube	.5PlogP	logP	logP

thousands of processors. For the time being, crossbars are restricted to small systems of at most 64 processors, or to providing interconnects among the processors of sub-clusters within larger machines.

Bus-based connection networks are attractive for moderate numbers of processors, for example 16 to 32. Beyond this point bandwidth begins to suffer intolerably. Architectures based on busses therefore tend to be hierarchial beyond that number of processors. As an example, the SUPRENUM-1 computer uses a fast local bus to connect within a cluster of 16 processors. Clusters are arranged in a rectangular grid and connected by row and column busses, which has the added attraction of providing redundancy and double bandwidth. New configurations of processors continue to be proposed. Of particular interest are Giloi and Montenegro's TICNET architecture (Giloi and Montenegro [5]), and Faber's vertex-symmetric minimal path networks (Faber and Moore [2]).

One recent trend is the move towards "worm-hole" routing in distributed systems. The basic idea here is to allow virtual circuits to be established between remote processors, and without the necessity of interrupting any intermediate nodes. While there may be a small overhead for circuit creation, subsequently all data traverses the circuit without overheads such as multiple startup costs at intermediate nodes. Once a circuit is established, communication proceeds essentially in bit-serial fashion. Frequently it suffices to create logical rather than physical connections. These allow messages to proceed on virtual worm-hole channels, but with the possibility that physically the channels are multiplexed. This is particularly convenient as a means for preventing dead-locks and blocking of small messages by large ones. The resulting communication performance tends to be essentially independent of distance. Worm-hole routing is utilized in the CM-2, the iPSC2, iWarp and the Symult among others. In the case of the Symult, the designers were so confident of the advantages of worm-hole routing that they abandoned a hypercube architecture from their first generation in favor of a simple two-dimensional rectangular grid.

2.2.5   Software

Software for currently available parallel computers is extremely limited. In all cases manufacturers provide Fortran and C compilers, which are frequently just a single-node processor compiler. These compilers have no concept of parallelism or of communication capability. Typical examples are the systems supplied by Intel, Symult and NCube. In these systems, all communication and process control is initiated explicitly by the user, resulting in substantial code modification as well as a loss of portability of software. Typically libraries of low-level communication primitives are supplied with these systems to allow the user to initiate communications operations. The resulting software is best described as "programming in communication assembly language".

A few manufacturers have gone beyond this step by providing language extensions that capture aspects of the parallel hardware. Thinking Machines provides a parallel Fortran for their Connection Machine CM-2 computer. The compiler supports the Fortran 8X array extensions to Fortran 77, and the convention is that objects declared as arrays are understood to be distributed across the parallel processors. Communication among processors is supported by the 8X shift operations, as well as the various reduction operators such as vector sum. While the Connection Machine programming environment is remarkably elegant and user-friendly, one should point out that the task is much simplified by the SIMD nature of the hardware which maps extremely well onto array operations.

Myrias Corporation and Evans and Sutherland both support a virtual address space across processors. If a processor attempts to access a memory location not in its physical memory, then a page fault occurs and the appropriate memory page is fetched from the processor that has it. Myrias in particular have implemented a sophisticated mechanism for load balancing and rapid access to memory. The system attempts to localize page table information and to provide access to it in a distributed fashion.

SUPRENUM supports extensions to Fortran for task control, and to assist in communication operations. In addition SUPRENUM is unique in providing a sophisticated high-level interface to the communication system. The library supports a range of grid-oriented operations that largely shield a numerical user from dealing with the communication system directly. In addition to providing powerful programming tools, such systems deliver the possibility of substantial program portability across architectures that support the common set of primitives.

One should also note the tendency to support virtual processes. This is an important aid to software development as it allows an application to simulate a larger number of processors than are physically present. Virtual processing is supported by the majority of systems in one form or another. Examples include iPSC, SUPRENUM, Symult and CM-2.

2.3 Some Representative Parallel Systems

In this section we will look briefly at the characteristics of a number of these machines. The machines currently under development have processor numbers ranging between 2 and 65,536. The machines listed above vary greatly in local processing power, ranging from a few megaflops up to 20 Gflops.

2.3.1 Evans and Sutherland ES-1

The ES-1 is a new (1989) parallel architecture based on a hierarchial crossbar structure. The basic "processor" is a complex package consisting of a cluster of 16 processing elements called computational units, along with 256 Mbytes of shared memory. Each processor supports an I/O subsystem with a bandwith of 160 Mbytes/sec over eight full duplex channels. Up to 8 processors and 8 I/O subsystems are currently supported. As mentioned previously, the ES-1 supports a virtual memory address space of 32 bits, which greatly simplifies program development for the system.

The processors and I/O subsystems are connected together by a crossbar. Each node (computational unit) is a 20 Mips 10 Mflops (64-bit precision) scalar processor with six pipelined functional units. The 16 nodes in a processor are connected by a 1 Gbyte/sec interconnect crossbar. The memory is 64-way interleaved and supports a bandwidth of 640 Mbytes/sec. The ES-1 operating system is a full Unix system and runs in every computational unit. Unlike most distributed processors, there is no front end processor. Since each cluster delivers 160 Mflops double precision, the most powerful current system delivers a peak rate of 1280 Mflops scalar and supports 2 Gbytes of memory.

2.3.2 CMU iWarp

The iWarp computer (Borkar *et al.* [1]) is a follow-on to the 100 Mflop Warp processor developed at Carnegie-Mellon University. The key advance in the iWarp is the development of a single chip processor combining the following functions: 20 Mflops computational power, 320 Mbyte/sec memory throughput and a communication engine with a latency of only 150 nanoseconds. The processor has been implemented as a 600,000 transistor custom

VLSI chip fabricated by Intel Corporation, hence the i in the name iWarp. Up to 64 Mbytes of memory is accessible per processor.

One important point is that the processor accomplishes 20 Mflops without pipelining. The adder unit delivers 5 Mflops (64-bit) or 10 Mflops (32-bit), non-pipelined, as does the multiplier unit. In addition the integer/logical unit delivers 20 Mips. All three units may perform simultaneously.

The system has been designed for flexibility from the start, and can be used efficiently to represent either general purpose distributed memory computers, or special purpose systolic arrays. The initial iWarp is an 8×8 array of processors delivering 1.2 Gflops, and expected to be available in 1990, but there are plans to extend this up to 1,024 processors.

One of the advances made in the iWarp project is the development of parallel program generators. These are tied to specific application domains — for example there is one for domain-based scientific computing, and another for image processing.

The communication facilities of iWarp are based on four input and output ports, each running at 40 Mbytes/sec. An input port of one iWarp processor may be connected directly to the output port of another processor to form a point-to-point communication network. A natural arrangement is thus to create one and two dimensional grids of processors. Because the communication processor performs independently of the numeric processor, worm-hole routing can be supported. Logical channels are supported by multiplexing of the physical communication lines, allowing for deadlock to be broken, and for long messages to be interrupted in worm-hole routing.

2.3.3   Connection Machine

The Connection Machine CM-1 designed by Thinking Machines, Inc., of Cambridge, MA, has 65,536 1-bit processors, though this may be regarded as a prototype for a machine that might have 1,000,000 processors. While designed primarily for artificial intelligence work, this machine has proved to have even greater potential applications to scientific computing applications (McBryan [6,8]). The more recent CM-2 computer adds 2,048 Weitek floating point co-processors and 512 Mbytes of memory, to provide a powerful computer for numerical as well as symbolic computing. The CM computers are SIMD machines. Logic is supported by allowing individual processors to skip the execution of any instruction, based on the setting of a flag in their local memory. The CM machines are based on a hypercube communication network, with a total communication bandwidth of order 3 Gbytes/sec. Communication is by worm-hole type routing. The system supports I/O to disks at up to 320 Mbyte/sec, and to frame buffers at 40 Mbyte/sec.

Connection Machine software consists of parallel versions of Fortran, C and Lisp. In each case it is possible to declare parallel variables, which are automatically allocated on the hypercube. Programs execute on a front-end machine, but when instructions are encountered involving parallel variables, they are executed as parallel instructions on the hypercube. The system supports the concept of virtual processors. A user can specify that he would like to compute with a million (or more) virtual processors, and such processors are then similar to physical processors in all respects except speed and memory size. A typical use is to assign one virtual processor per grid point in a discretization application. This provides a very convenient programming model. Parallel global memory reference is supported using both regular multi-dimensional grid notations (NEWS communication) and random access (hypercube) modes.

2.3.4 Myrias

The Myrias SPS-2 computer, built by Myrias Research Corp. of Edmonton, Alberta, has up to 1024 processing elements. The architecture is a hierarchial bus design, utilizing 33 Mbytes/sec busses to interconnect processors within clusters and clusters to each other. Each processor is a 32-bit Motorola 68020 microprocessor with 4 Mbytes of local memory. The architecture is a three-level hierarchial system. Processors are assembled in groups of four on a board connected among each other by a bus, along with an I/O port controller. At the second level in the hierarchy is the card cage, containing 16 processor boards and thus 64 processors, as well as one or two off-cage communication boards. Each communication board supports four off-cage links which can be connected to other cages or to the front end computer.

The SPS-2 supports a global 32-bit virtual address space. There is no concept of shared access to memory locations. Simple extensions of Fortran support parallel do and a join operation. The Par Do model used by Myrias is somewhat unusual in that there are no possibilities for sharing data. A Par Do is executed by specifying a code segment to be executed and the number of child tasks to be run. Each thread of execution performs quite independently in its own address space, starting with a copy of the parents memory. Execution of a child proceeds in normal sequential mode, except that Par Do's may be nested recursively. On completion of all children, the memory states of the children are merged to form the new memory state of the parent. Thus a child can never affect the memory of another child, but can affect the memory state of its parent, but only after all children merge.

The rules for merging of child memories on completion are:

1. If no child stored a value at the address, the location in the parent

memory retains its original value.

2. If exactly one child changed a value at the address, the location in the parent receives the last value from the child.

3. If more than one child stores a value at the address the result is unpredicatable unless all values stored are the same. Efficiency is maintained throughout the process by using a copy-on-write approach which ensures that most of the global address space is never really replicated.

The programming approach to the system is thus functional in nature and is based on parallel recursion. As an example, we describe an implementation of a global vector sum, assuming the vector is initially of length N. The parent creates 2 tasks, each intended to sum half of the elements, storing the results in variables left and right. Each child task similarly spawns two more children, storing results into 2 local variables of the principal child, and so on. On completion of a pair of child tasks, its parent has the two partial sums in separate variables, and therefore it proceeds to add these, providing its value to the next parent above. In this way the result is obtained at the global parent in time log(N) without ever sharing memory. All assignment of work to processors is handled automatically.

2.3.5   SUPRENUM

The German SUPRENUM project involves coupling up to 16 processor clusters with a network of 200 Mbyte busses. The busses are arranged as a rectangular grid with 4 horizontal and 4 vertical busses. Each cluster consists of 16 processors connected by a fast bus, along with I/O devices for communication to the global bus grid and to disk and host computers. There is a dedicated disk for each cluster. Individual processors can deliver up to 16 Mflops of computing power and support 8 Mbytes of memory. The very high speed of the bus network makes this an interesting machine for a wide range of applications, including those requiring long-range communication. No more than three communication steps are ever required between remote nodes. A prototype cluster containing 16 nodes is already in operation, and a full machine with up to 16 clusters will be available in late 1989.

SUPRENUM is characterized by the best support for scientific applications to be found among the various vendors. The effort invested in development of libraries of high-level grid and communication primitives will greatly ease the effort of moving applications to the computer, and also provides substantial high-level portability to other systems, since the communication library can be implemented in terms of low level primitives on any distributed system.

2.3.6   Intel iPSC

The Intel iPSC was the first commercial hypercube computer, and has been the most widely available highly parallel computer in recent years. Built from 128 Intel 80286 processors, peak computer power was under 10 Mflops, yet the iPSC was the basis for a large number of useful experiments in parallel computing. The recently developed iPSC/2 computer is a second generation machine that provides greatly increased processing power and communication throughput. Each node contains an 80386 microprocessor with up to 8 Mbytes of memory (extendible to 16Mbytes with 64 processors). There are three available numeric co-processors: an Intel 80387 co-processor (300 Kflops), a Weitek 1167 scalar processor (900 Kflops) and a VX vector board (6 Mflops double precision, maximum of 64 nodes). Thus the top-rated system has 64 nodes capable of 424 Mflops double precision and 1280 Mflops single precision. Special communication processors allow message circuits to be established between remote processors without intervention from intermediate processors. Thus the iPSC now implements worm-hole routing rather than the store and forward protocol of previous generations.

2.3.7   Symult

The Symult 2010 is a new commercial machine based on a grid architecture. Individual nodes consist of a Motorola 68020 (25 MHz) with a 68881 co-processor (150 Kflops). An optional upgrade to the 68882 processor (215 Kflops) is possible. A further option provides for a vector processing board based on Weitek chips, with a 20 Mflops performance. Peak performance of a 1024 node system may be as high as 20 Gflops. Memory per node ranges from 2 to 10 Mbytes. Maximum node memory is 8 Mbytes, with a further 10 Mbytes of memory on the vector board. The most interesting feature of the machine is the message routing system which establishes point-to-point communications between remote nodes. Each node has a routing device that can support simultaneous transmission on four links at 20Mbytes each, without interruption of intermediate computational processors. Communication is by "worm-hole" connectivity rather than the usual store-and-forward, resulting in far greater performance for long-range communication. In this communication mode, a connection circuit is first established between remote nodes, incurring a small startup cost, after which the message is transferred in bit-serial fashion in a single operation. Worm-hole communication provides extremely fast long-distance communication, whereas a standard store and forward model would incur large overheads due to the long path-lengths on a grid.

2.3.8   AMT DAP

The DAP was the first massively parallel single-bit computer, and has been widely used for a range of scientific applications. Its current incarnation as the AMT 510 attached processor, provides the capability to attach a 1024 processor DAP array to any VAX or SUN computer. The 510 is a 32x32 array of processors, arranged as a two-dimensional grid and is implemented in VLSI on 16 chips. Additional busses connect all processors on each row and column and are used for broadcasts and other non-local operations. Up to 1 Mbit of memory may be installed per processor, for a combined total of 128 Mbytes. The computer is SIMD, and can execute at up to 60 Mflops, although boolean operations perform at up to 10 Gips.

2.3.9   Paralex Gemini and Pegasus

Paralex Research Inc. is developing a line of highly parallel local memory systems in the supercomputer class. The initial Gemini product supports up to 1000 nodes with peak performance up to about 2 Gips and 500 Mflops. The Gemini uses a hypercube to provide connectivity, and also features a high performance UNIX front end. The second generation Pegasus machine, due in 1989, will support 512 nodes with 8 Gbytes of memory and will provide 25 Gips and 15 Gflops peak rate. This system will be based on the new SPARC technology being licensed by SUN Microsystems. The follow-on Genesis system, planned for 1990, will provide up to 2 Tflops (teraflops) of performance.

2.3.10   GF-11 and TF1

The GF-11 is an IBM parallel computer, designed to perform very specific scientific computations at Gflop rates. The GF-11 has 576 processors (including 64 backup processors), coupled through a three stage Benes network which can be reconfigured at every cycle in 1024 different ways by an IBM 3084 control processor. Peak processing power of 11 Gflops will allow previously uncharted computational regimes to be explored. The machine has been designed primarily for solving quantum field theory problems and is not a general purpose computer; in particular, very little software is available. It is an SIMD architecture but with some flexibility in that the settings of local registers may be used to control the behavior of individual processors.

2.3.11 CCI Navier-Stokes Machine

The NASA-sponsored Navier-Stokes Machine being built at Princeton University involves an experiment with reconfigurable pipelines as well as parallelism. Up to 64 processors are supported with hypercube connections. Each node consists of a CPU, 32 arithmetic processing units and 2 Gbytes of memory. Each of the arithmetic units may be specified to be an adder, multiplier, etc., and connections can then be specified between them in order to represent efficiently a pipeline to evaluate an expression. Reconfiguring the connections takes only 50 nano-seconds. Since each arithmetic unit has a peak processing power of 20 Mflops, the combined processing power per node is 640 Mflops. CCI Corporation plans to market a commercial version of the Navier-Stokes Machine.

2.3.12 HEP and TERA

The Denelcor HEP was the first commercial parallel computer. The HEP featured a shared memory, with special access bits to provide for memory locking on every word. The processors were pipelined units, each capable of executing a large number of separate instruction streams simultaneously. Each processor was rated at 10 Mips. The new TERA computer, designed by HEP creator Burton Smith, will support 256 processors, each similar in many respects to the HEP, and will provide up to 256 Gflops of computational power in a shared memory, scalar processing environment.

2.3.13 Other Approaches

A variety of other important architectures are also under development. These include various dataflow machines (with bus, tree and grid structures), examples include the MIT Tagged Token machine, the NTT Dataflow grid machine and the Manchester Dataflow Machine. Another important class are the tree-structured machines (binary trees, trees with sibling or perfect shuffle connections), examples of which are the Columbia University DADO machine and the CMU Tree Machine. Because of the simplicity of the connections, nearest neighbor machines, such as the MPP, and ring architectures, such as the University of Maryland's ZMOB (256 processors on a ring), are also popular designs.

References

[1] Borkar, S., R. Cohn, G. Cox, S. Gleason, T. Gross, H. T. Kung, M. Lam, B. Moore, C. Peterson, J. Pieper, L. Rankin, P. Tseng, J. Sutton, J. Ur-

banski, and J. Webb, "iWarp: An Integrated Solution to High Speed Parallel Computing," in *Proceedings of Supercomputing '88 Conference*, Orlando, Florida, 330–339, Nov. 1988.

[2] Faber, V., and J. Moore, *High-degree Low-Diameter Interconnection Networks with Vertex Symmetry: The Directed Case*, Los Alamos Technical Report LA-UR-88-1051, March 1988.

[3] Flynn, M. J., "Very High-Speed Computing," *Proc. IEEE*, 54, 1901–1909, 1966.

[4] Frederickson, P. O., and O. McBryan, "Parallel Superconvergent Multigrid," in *Multigrid Methods: Theory, Applications and Supercomputing*, S. McCormick (ed.), Math Applications Series, Marcel-Dekker Inc., New York, 110, 195–210, 1988.

[5] Giloi, W. K., and S. Montenegro, *Super Interconnection Networks for Super Computers*, GMD Technical Report, Berlin, 1988.

[6] McBryan, O., "New Architectures: Performance Highlights and New Algorithms," *Parallel Computing*, 7, 477–499, 1988.

[7] McBryan, O., *Solving PDE at 3.8 Gigaflops*, University of Colorado CS Dept., Sept 1987.

[8] McBryan, O., "The Connection Machine: PDE Solution on 65536 Processors," *Parallel Computing*, 9, 1–24, 1988.

[9] McBryan, O., and E. Van de Velde, "Hypercube Algorithms and Implementations," *SIAM J. Sci. Stat. Comput.*, 8, 2, 227–287, 1987.

[10] McBryan, O., and E. Van de Velde, "Parallel Algorithms for Elliptic Equations," *Commun. Pure and Appl. Math.*, i 38, 769–795, 1985.

[11] McBryan, O., and E. Van de Velde, "The Multigrid Method on Parallel Computers," in *Proceedings of 2nd European Multigrid Conference*, Cologne, J. Linden (ed.), GMD Studie Nr. 110, GMD, July 1986.

[12] Schwartz, J., *A Taxonomic Table of Parallel Computers, Based on 55 Designs*, Ultracomputer Note #69, Courant Institute, New York, 1983.

Chapter 3

Machine Independent
Parallel Numerical Algorithms

Robert A. van de Geijn[*]

3.1 Introduction

In this chapter, we explore an approach to programming parallel matrix algorithms that takes advantage of the inherent structure of the types of communications encountered in such algorithms to achieve the following goals:

1. Ease of programming: Experience has shown that restricting the types of communication to a limited set of frequently encountered communication primitives allows straight-forward parallel implementations of typical matrix algorithms.

2. Portability: By coding parallel algorithms in terms of a few communication primitives, the code can be ported to any parallel processor on which these primitives have been implemented.

3. Reduced communication overhead: The communication primitives can be optimized for a specific parallel architecture. By restricting the types of communication, the parallel environment can be simpler and is therefore likely to incur lower communication overhead.

---

[*]Department of Computer Sciences, The University of Texas, Austin, TX.

4. Reasonable efficiency: Our goal is to create an environment that balances the efficiency achieved with the effort required to reach this efficiency. In other words, we wish to create an environment that allows the user to achieve a reasonable speedup by using the parallel processor without spending an unreasonable amount of time fine-tuning the code for the specific parallel architecture.

For the moment, we restrict the class of parallel processors for which the parallel environment is intended to parallel processors which meet the following requirements:

1. The parallel processor must be an MIMD machine (i.e. each processor has its own CPU) with an adequate amount of memory (e.g. at least $\frac{1}{2}$Mbyte).

2. Communication must be "reasonably" fast. We discuss what we consider reasonable in a later section.

3. The number of processors is restricted to fewer than about 1000.

It should be noted that a wide range of currently available parallel processors meet these criteria, including the BBN Butterfly, the Sequent Balance, the Intel iPSC2, NCUBE Inc.'s NCUBE, and the AMETEK 2010.

We present the primitives to which we intend to restrict communication in Section 3.2. An example showing how these primitives can be used to program matrix algorithms is presented in Section 3.3, while other applications are mentioned in Section 3.4. Limitations and extensions of this approach to programming parallel matrix algorithms are discussed in Section 3.5.

We cannot take credit for the observation that the primitives mentioned in Section 3.2 occur frequently in parallel implementation of numerical algorithms. The theoretical analysis of implementing a similar set of primitives on different parallel architectures was studied by Saad and Schultz [6]. Our approach is unique in that it recognizes the advantages of restricting communication to only these primitives.

3.2 Parallel Communication Primitives

In this section, we present a set of communication primitives that we believe suffices to implement matrix algorithms efficiently on the type of parallel processors mentioned in the introduction. We will give an indication of how time consuming the various communication patterns are by concentrating on the popular hypercube architecture. For reasons that will become apparent in Section 3.5, we use the term node instead of processor in the next sections.

A hypercube of dimension D consists of $p = 2^D$ nodes, P_0, ..., P_{p-1}. Nodes P_i and P_j are neighbors if and only if the binary representations of i and j differ in exactly one binary digit. One important property of a hypercube is that it is possible to embed a ring, i.e., there exists a permutation Π such that if $(\Pi(i) - \Pi(j))$ mod $p = 1$, then P_i and P_j are neighbors.

We will assume that sending a packet of m data items (e.g. floating point numbers) between neighboring nodes requires time $\alpha + m\beta$, where α represents the startup time. A floating point operation requires time γ. Furthermore, all nodes can send the same data to different neighbors simultaneously and all links between nodes are bidirectional. That is, data can be sent in both directions simultaneously. The time complexities given are not necessarily optimal.

The primitives are:

1. Node-to-neighbor: One node sends a packet of m items to a neighboring node (within the embedded ring). Time complexity on the hypercube is $\alpha + m\beta$.

2. Broadcast: One node sends a packet of m items to all other nodes. Time complexity on the hypercube (Ipsen *et al.* [2]) is

$$\left(\sqrt{\log_2(p)\alpha} + \sqrt{m\beta}\right)^2 \quad \text{if} \quad 1 \leq \frac{m}{\log_2(p)}\frac{\alpha}{\beta} \leq m^2\,,$$
$$(\log_2(p) + m)(\alpha + \beta) \quad \text{if} \quad \frac{m}{\log_2(p)}\frac{\alpha}{\beta} < 1\,,$$
$$\log_2(p)(\alpha + m\beta) \quad\quad \text{otherwise}\,.$$

For the sake of argument, we will use the time complexity $(\sqrt{\log_2(p)\alpha} + \sqrt{m\beta})^2$ and bound it by $2\log_2(p)\alpha + 2m\beta$.

3. Total exchange: All nodes send a packet of m items to all other nodes. Time complexity on the hypercube is $\log_2(p)\alpha + pm\beta$.

4. Data transpose: Each node P_i has p packets, each of size h, denoted by x_{ij}, $0 \leq j < p$. Let $m = ph$. After the data transpose, each node P_i has packets x_{ji}, $0 \leq j < p$. Time complexity on the hypercube is $\log_2(p)(\alpha + \frac{m}{2}\beta)$.

5. Left/right-shift: Each node sends a packet of m items to the node to its left/right (within the embedded ring). Time complexity on the hypercube is $\alpha + m\beta$.

6. Vector-sum: Each node P_i owns a vector of length $n = hp$, where h is an integer. After the vector-sum, the result of summing all p vectors is

distributed among the nodes, i.e. each node has a part of length h of the result. Time complexity on the hypercube is $\log_2(p)\alpha + n\beta + n\gamma$.

7. Inner-product: Let x, y be vectors of length $n = hp$. Assume each node has parts of x and y of length h. The inner-product primitive computes $x^T y$, leaving the result at each node. Time complexity on the hypercube is $h\gamma + \log_2(p)(\alpha + \beta + \gamma)$.

8. Global compare: Assuming each node has a number z_i, the global compare determines the largest number, distributing it to all nodes, together with the index of the node that originally owned it. Time complexity on the hypercube is $\log_2(p)(\alpha + 2\beta)$.

In the next section, an example that utilizes several of the above primitives is given.

3.3   Example: Reduction to Upper Hessenberg Form

Given matrix $A \in I\!\!R^{n \times n}$, in a typical implementation of the QR methods for finding the eigenvalues of A, the matrix is first reduced to upper Hessenberg form using unitary similarity transformations. We show how the broadcast, vector add, and total exchange operations can be used to parallelize a standard algorithm for this reduction. We chose this particular example because it encompasses a wide range of computations and communications encountered in typical parallel matrix algorithms.

Partition $x \in I\!\!R^n$ so that $x^T = (x_1^T, x_2^T)$, where $x_1 \in I\!\!R^i$ and $x_2 \in I\!\!R^{n-i}$. Define $P_i(x) = I - 2uu^T$, where $u = v/\|v\|_2$ and $v^T = (x_1^T, x_2^T) \pm \|x_2\|_2 e_{i+1}^T$. This is an example of a Householder transformation. Note that $P_i(x)x = (x_1^T, \pm\|x_2\|_2, 0, \ldots, 0)^T$.

Let $A^{(i)} = (a_1^{(i)}, \ldots, a_n^{(i)})$ be a column partitioning of $A^{(i)}$. The following algorithm successively annihilates the elements of A that lie below the first subdiagonal:

Algorithm 3.1 Reduction to upper Hessenberg form.

$$A^{(0)} = A$$
$$\text{for } k = 1, \ldots, n-2$$
$$\quad \text{compute } P_k = P_k(a_k^{(k-1)}) = I - 2uu^T$$
$$\quad A \leftarrow A^{(k)} = P_k A^{(k-1)} P_k^T$$

Turning to the parallel implementation of this algorithm, let a_j be assigned to node $P_{(j-1)\bmod p}$. This storage scheme is called column-wrapped storage. We will use the notation $j \in P_i$ to indicate that column j resides on P_i.

Assume the computation has progressed to the point where $A = A^{(k-1)}$. The node that owns $a_k^{(k-1)}$ can compute u such that $P_k = P_k(a_k^{(k-1)}) = I - 2uu^{\mathrm{T}}$. First, we must form $A^+ = P_k A^{(k-1)} = A - 2uu^{\mathrm{T}}A$. Note that this is equivalent to updating $a_j^{(k-1)}$ with $a_j^+ = a_j^{(k-1)} - 2u^{\mathrm{T}}a_j^{(k-1)}u$, $k < j \le n$. Hence, once u has been distributed to all nodes, each node P_i can update $a_j^{(k-1)}$, for all $j \in P_i > k$.

Next, we must form $A^{(k)} = A^+ - 2A^+uu^{\mathrm{T}} = A^+ - 2yu^{\mathrm{T}}$, where $y = \sum_{j>k} u_j a_j^+$. The vector y can be formed by simultaneously computing $y^{(i)} = \sum_{j\in P_i>k} u_j a_j^+$, performing a parallel vector-add, which leaves $y = \sum_{i=0}^{p-1} y^{(i)}$ distributed among the nodes, followed by a total exchange, which distributes y to all nodes. Finally, each node P_i can form $a_j^{(k)} = a_j^+ - 2u_j y$, for all $j \in P_i > k$. Pseudo code driving node P_i is given by Algorithm 3.2.

Algorithm 3.2 Parallel Reduction to upper Hessenberg form.

```
for k = 1, ..., n − 2
    if (k − 1) mod p = i
        compute P_k = P_k(a_k) = I − 2uu^T        [(n − k)γ]
        broadcast u                                [2log₂(p)α + 2(n − k)β]
    else receive u
    update
        a_j ← a_j − 2u^T a_j u,  j ∈ P_i > k       [2[(n−k)/p](n − k)γ]
        compute y^(i) = Σ u_j a_j,  j ∈ P_i > k    [[(n−k)/p]nγ]
        vector-sum y = Σ y^(j) leaving y_i on P_i  [log₂(p)α + nβ + nγ]
        total exchange y_i                         [log₂(p)α + nβ]
        set a_j = a_j − 2u_j y,  j ∈ P_i > k       [[(n−k)/p]nγ]
```

The expressions in brackets indicate the contribution to the total time complexity of the computation.

The total time complexity on a hypercube equals

$$\sum_{k=1}^{n-2} \left(2n - k + 2\lceil \frac{n-k}{p} \rceil (n-k+n) \right) \gamma + 4\log_2(p)\alpha + (4n - 2k)\beta.$$

Bounding $\lceil x \rceil$ by $x + 1$, we find that the time complexity is approximated by

$$T_p(n, \alpha, \beta, \gamma) = \left(\frac{5}{3}\frac{n^3}{p} + 4n^2 \right) \gamma + 4n\log_2(p)\alpha + 3n^2\beta,$$

ignoring lower order terms. This compares favorably with the sequential time complexity of $T_1(n) = \frac{5}{3}n^3\gamma$. In fact, if p is fixed and the problem size increases, 100% utilization of the nodes will be approached.

We define the efficiency of the parallel algorithm by

$$E_p(n) = \frac{T_1(n)}{pT_p(n,\alpha,\beta,\gamma)} \quad .$$

Clearly the efficiency is affected by the constants α, β, and γ. A typical current generation hypercube has $\alpha \approx 250\mu$sec and $\beta \approx 10\mu$sec, compared to $\gamma \approx 10\mu$sec. The efficiency of our algorithm on such machines is indicated in Figure 3.1. The column-wrapped storage requires at least one column per node. We chose to start our plots when at least two columns were assigned per node. Reasonable efficiencies can be obtained, even when p and n are of the same order.

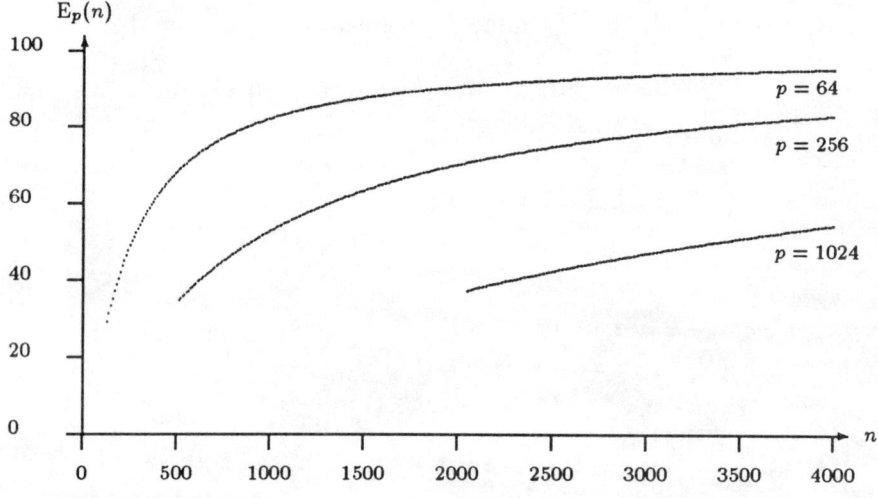

Figure 3.1: Projected efficiency of the reduction to upper Hessenberg form when $\alpha = 250\mu$sec, $\beta = 10\mu$sec, and $\gamma = 10\mu$sec.

3.4   Other Applications

To give an indication of the range of algorithms that can be implemented using this approach, in this section we give a small but representative sample of other matrix algorithms and the primitives they require.

Solution of Dense Linear Systems

Assuming the matrix is distributed using column-wrapped storage, Gaussian elimination requires only a broadcast primitive: To eliminate the entries of the kth column that are below the diagonal, the processor that owns the kth column computes the multipliers and broadcasts them. Next, all processors update their part of the columns with index greater than k. For details, see Ipsen *et al.* [2] and Juszczak and Van de Geijn [3].

For parallel implementation of an algorithm for solving the resulting triangular systems, we refer the reader to Li and Coleman [4], who present an implementation that requires only node-to-neighbor (within the embedded ring) communication.

Eigenvalues of Dense Matrices

Parallel implementation of Jacobi's method and the nonsymmetric QR algorithm for the symmetric and nonsymmetric eigenvalue problems, respectively, require the broadcast, total exchange and left/right shift primitives. Implementations that use a storage scheme other than column-wrapped storage can be found in Van de Geijn [7] and Van de Geijn and Hudson [8].

Solutions of Sparse Linear Systems

A simple conjugate-gradient method, without preconditioning, can be implemented in parallel using the inner-product and left/right shift primitives.

For our brief description, we concentrate on the model problem

$$\frac{\partial^2 u}{\partial x^2} + \frac{\partial^2 u}{\partial y^2} = 0, \quad 0 < x, y < 1,$$

on a rectangular mesh that has been ordered as in Figure 3.2. We assume that lines of mesh points are assigned to processors in a wrapped fashion so that line k resides on $P_{k \bmod p}$ and will use the notation $j \in P_i$ to indicate mesh point value u_j resides on processor P_i. The discretization leads to the linear system $Au = b$, where the matrix is blocked, $A = (A_{ij})$, with $A_{ij} \in R^{n \times n}$,

$$A_{ii} = \begin{pmatrix} 4 & -1 & & \\ -1 & 4 & -1 & \\ & -1 & 4 & \ddots \\ & & \ddots & \ddots \end{pmatrix},$$

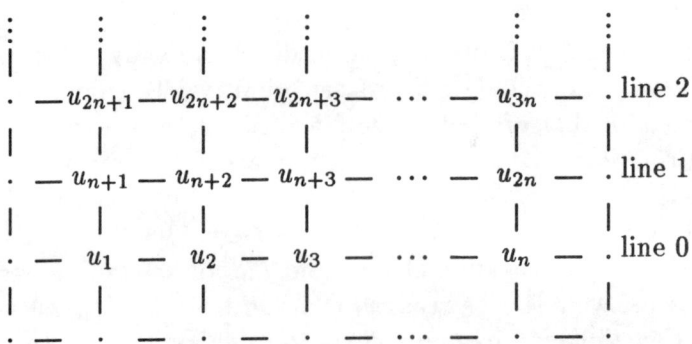

Figure 3.2: Discretization mesh for model problem.

$A_{i(i+1)} = A_{(i+1)i} = -I$, and zeroes elsewhere.

A conjugate-gradient method for solving this system is given by (Golub and Van Loan [1]):

Algorithm 3.3 Conjugate-gradient method.

$$u = 0; \; r = b; \; \rho_0 = r^{\mathrm{T}} r; \; k = 1$$
Do until convergence
if $k = 1$ then $p = r$
else $\beta = \rho_{k-1}/\rho_{k-2}; \; p = r + \beta p$
$$w = Ap; \; \alpha = \rho_{k-1}/p^{\mathrm{T}} w; \; u = u + \alpha p;$$
$$r = r - \alpha w; \; \rho_k = r^{\mathrm{T}} r; \; k = k + 1$$

Turning to the parallel implementation, let all vectors be assigned to processors in the same manner as u. Algorithm 3.3 can be easily changed to pseudo code driving P_i by observing that each statement that updates a vector, e.g. $p = r + \beta p$, can be performed in parallel by letting P_i update only those components of the vector that are are assigned to its memory,

$$p_j = r_j + \beta p_j, \quad \forall j \in P_i .$$

Each inner-product that is performed can be replaced by a call to the parallel inner-product primitive, which leaves the result at all processors. Any scalar operation is simply duplicated by all processors. At first, it may appear that the operation $w = Ap$ requires a parallel vector-add primitive, similar to the formation of A^+u in the parallel reduction to upper Hessenberg form.

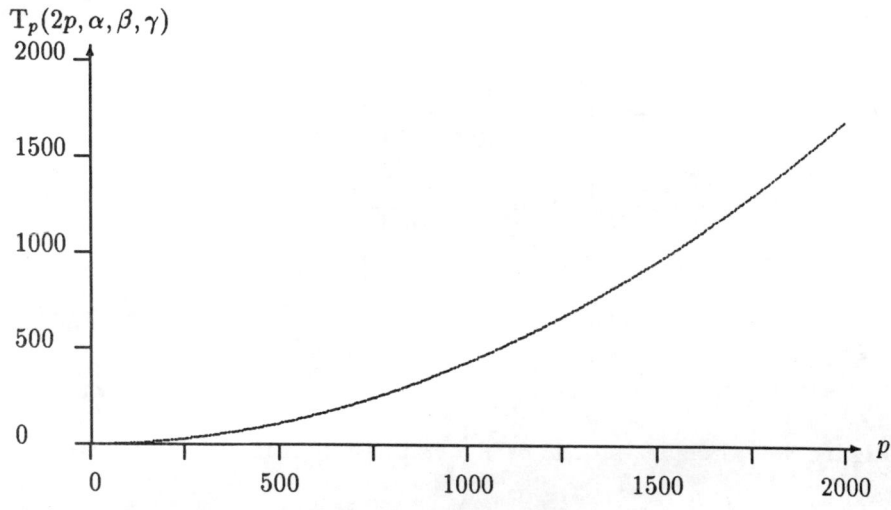

$T_p(2p, \alpha, \beta, \gamma)$

Figure 3.3: Projected computation time (in sec.) when $n = 2p$.

However for the model problem, a typical component of w is $w_j = 4p_j - p_{j-1} - p_{j+1} - p_{j-n} - p_{j+n}$. While w_j, p_j, p_{j-1}, and p_{j+1} already reside on the same processor, p_{j-n} and p_{j+n} are owned by the left and right neighbor, respectively, and hence the formation of w merely requires p to be shifted both to the right and the left. In other words, $w = Ap$ is replaced by

```
left- and right-shift p
```
$$w_j = 4p_j - p_{j-1} - p_{j+1} - p_{j-n} - p_{j+n}, \quad \forall j \in P_i;$$

The statement must be slightly modified when a boundary of the mesh is involved.

To make the conjugate-gradient method more effective, a preconditioning scheme must be added. We are currently studying parallel implementation of this addition. Parallel implementations of multi-grid methods using techniques similar to ours can be found in McBryan and Van de Velde [5].

3.5 Limitations and Extensions

Let us again consider the time complexity of the parallel reduction to

$P_0 \quad P_1 \quad P_2 \quad P_3$

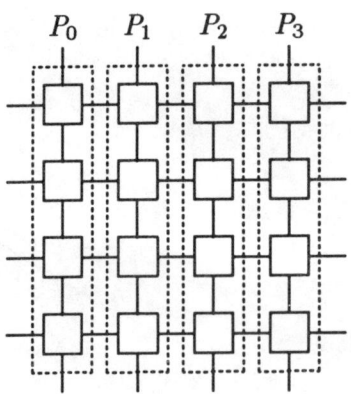

Figure 3.4: Grouping processors in a grid to form more powerful nodes.

upper Hessenberg form,

$$T_p(n, \alpha, \beta, \gamma) = \left(\frac{5}{3} \frac{n^3}{p} + 4n^2 \right) \gamma + 4n \log_2(p)\alpha + 3n^2\beta.$$

The term $4n^2\gamma$ is mostly attributable to the fact that nodes are idle while one node computes u as well as the fact that the computational load is not evenly distributed despite the wrapping of the columns. Notice that if n/p is small, this term is on the order of $\frac{5}{3}\frac{n^3}{p}\gamma$, which suggests that we can expect a loss of efficiency as p approaches n. Similarly, $3n^2\beta$ becomes significant as p approaches n. It was already noted that p cannot exceed n when column-wrapped storage is used.

As the number of nodes of the parallel processor increases, so must the size of the problem, n, when we restrict ourselves to column-wrapped storage. Figure 3.3 indicates the projected total computation time for the reduction to upper Hessenberg form when the number of nodes is increased while the size of the problem is increased proportionally, by restricting $n = 2p$. (The same constants for α, β, and γ were used.) We learn from this figure that eventually the problem assigned to the parallel processor must become so large that the computation time exceeds what we consider reasonable. For this reason, we recommend our current approach only for parallel processors with at most 1024 processors.

It would appear that if we wish to further increase computational power of the parallel processor, each individual node must be made more powerful and the penalty for communication must be decreased (i.e. β and γ must decrease, while keeping p constant). However, note that more powerful nodes can be created by grouping several processors, as illustrated in Figure 3.4. Columns of a matrix can then be wrapped onto the nodes as before. This leaves us to decide how to assign parts of the columns to the processors that make up each node. One possibility is to wrap the elements of each column onto the processors of the node that owns the column. We are currently investigating how to adapt the communication primitives to accommodate nodes which are created by grouping processors.

3.6 Conclusion

The parallel algorithm for reduction to upper Hessenberg form illustrates how a handful of communication primitives can be used to efficiently implement matrix algorithms on parallel processors. The complexity analysis shows that reasonable efficiency can be achieved even when communication and computation are separated.

We are currently developing a communication library based on these primitives. The theoretical results will be verified by experiments on currently available parallel processors, including the AMETEK 2010, the Sequent Balance, and BBN's Butterfly. We intend to show that a wide range of algorithms can be implemented using the communication library by concentrating on standard routines in packages like LINPACK and EISPACK. Other future research includes extending the primitives to parallel processors with larger numbers of processors by applying the techniques mentioned in Section 3.5.

References

[1] Golub, G.H. and C.F. Van Loan, *Matrix Computations*, Johns Hopkins Press, 1983.

[2] Ipsen, I.C.F., Y. Saad, and M.H. Schultz, "Complexity of Dense-Linear-System Solution on a Multiprocessor Ring," *Linear Algebra and its Applications*, 77, 205-239, 1986.

[3] Juszczak, J.W. and R.A. van de Geijn, "An Experiment in Coding Portable Parallel Matrix Algorithms," in *Proceedings of the Fourth Annual Hypercube Conference*, to appear, 1989.

[4] Li, G. and T.F. Coleman, "A Parallel Triangular Solver for a Distributed-Memory Multiprocessor," *SIAM J. Sci. Stat. Comput.*, 9, 3, 382–396, 1988.

[5] McBryan, O.A. and E.F. Van de Velde, "Hypercube Algorithms and Implementations," *SIAM J. Sci. Stat. Comp.*, 8, 2, s227–s287, 1987.

[6] Saad, Y. and M.H. Schultz, *Data Communication in Parallel Architectures*, Research Report YALEU/DCS/RR-461, 1986.

[7] Van de Geijn, R. A., "Storage Schemes for Parallel Eigenvalue Algorithms," in *Numerical Linear Algebra, Digital Signal Processing and Parallel Algorithms*, G. Golub and P. Van Dooren (eds.), NATO ASI Series, Springer Verlag, to appear, 1989.

[8] Van de Geijn, R.A. and D.G. Hudson III, "Efficient Parallel Implementation of the Nonsymmetric QR Algorithm," in *Proceedings of the Fourth Annual Hypercube Conference*, to appear, 1989.

Chapter 4

A Linear Algebra Library
for High-Performance Computers

*Christian H. Bischof** and Jack J. Dongarra**

4.1   Introduction and Objectives

Argonne National Laboratory, the Courant Institute for Mathematical Sciences, and the Numerical Algorithms Group, Ltd., are developing a transportable linear algebra library in Fortran 77. The library is intended to provide a uniform set of subroutines to solve the most common linear algebra problems and to run efficiently on a wide range of high-performance computers.

The LAPACK library (shorthand for Linear Algebra Package) will provide routines for solving systems of simultaneous linear equations, least-squares solutions of overdetermined systems of equations, and eigenvalue problems. The associated matrix factorizations (LU, Cholesky, QR, SVD, Schur, generalized Schur) will also be provided, as will related computations such as reordering of the factorizations and condition numbers (or estimates thereof). Dense and banded matrices will be provided for, but not general sparse matrices. In all areas, similar functionality will be provided for real and complex matrices.

The new library will be based on the successful EISPACK [27,36] and LINPACK [11] libraries, integrating the two sets of algorithms into a unified,

---

*Mathematics and Computer Science Division, Argonne National Laboratory, Argonne, IL.

systematic library. A great deal of effort has also been expanded to incorporate design methodologies and algorithms that make the LAPACK codes more appropriate for today's high-performance architectures. The LINPACK and EISPACK codes were written in a fashion that, for the most part, ignored the cost of data movement. Most of today's high-performance machines, however, incorporate a memory hierarchy [20,31,38] to balance the difference in speed of memory accesses and vectorized floating-point operations. As a result, codes must be careful about reusing data in order not to run at memory speed instead of floating-point speed. LAPACK codes have been carefully restructured to reuse as much data as possible in order to reduce the cost of data movement. Further improvements are the incorporation of new and improved algorithms for the solution of eigenvalue problems [10,16].

LAPACK is designed to be efficient and transportable across a wide range of computing environments, with special emphasis on modern high-performance computers. While we do not hope for LAPACK codes to be optimal for all architectures, we expect high performance over a wide range of machines. By relying on the Basic Linear Algebra Subprograms (BLAS) [18,19,32] the codes can be "tuned" to a given architecture by efficient—and, in all likelihood machine-dependent—implementations of these kernels. Machine-specific optimizations are limited to those kernels, and the user interface is uniform across machines. We will also distribute test and timing routines to verify the installation of the LAPACK codes on a particular architecture and to allow for easy comparison with existing software.

Netlib [13] has demonstrated how useful and important it is for libraries to be easily available, and preferably on line. We intend to distribute the new library in a similar way, for no cost or a nominal cost only.

4.2   Use of the BLAS

The first BLAS [32], which we will call Level 1 BLAS from now on, implement common vector-vector operations such as a dot product or a "saxpy"

$$y \leftarrow y + ax.$$

The Level 2 BLAS [18] provide matrix-vector operations such as matrix-vector multiplication and rank-one updates. The development of the Level 2 BLAS was motivated by vector-processing machines. Many of the frequently used algorithms of numerical linear algebra can readily be coded so that the bulk of the computation is performed by calls to the Level 2 BLAS routines; efficiency can then be obtained by utilizing a tailored implementation of the Level 2 BLAS routines. On vector-processing machines, the aims of such an implementation are to keep the vector lengths as long as possible and to reuse results in vector registers.

Unfortunately, this approach is often not well suited to computers with a memory hierarchy (such as global memory, cache or local memory, and vector registers) and parallel-processing computers. (For a description of many advanced-computer architectures see [20,31,38].) Data at low levels of the memory hierarchy can be accessed without delay, whereas data at higher levels is available only after some delay and (because of memory bank conflicts) may not be available at a rate fast enough to feed the arithmetic units. For this reason it is imperative to reuse data as much as possible to cut down on data movement overhead.

This goal can be achieved by expressing a computation in terms of matrix-matrix operations. If the matrices involved are of order n, matrix-matrix operations such as matrix-matrix multiplication require $O(n^3)$ floating-point operations; with proper implementation, however, the data movement overhead is only $O(n^2)$. In contrast, vector-vector or matrix-vector operations each require $O(n)$ floating-point operations and $O(n^2)$ data movement overhead. Hence, using matrix-matrix operations, we avoid excessive movement of data to and from memory and achieve a *surface-to-volume effect* for the ratio of operations to data movement.

The Level 3 BLAS [12] provide the matrix-matrix operations needed for linear algebra. Together with the Level 1 and 2 BLAS, they provide a well-defined interface for the elementary matrix and vector operations and add to the portability, modularity, and ease of maintenance of the software. It is likely that the LAPACK project will reveal the need for a few additional basic routines whose performance may need to be optimized for different architectures and may be regarded as extensions to the current set of BLAS (e.g., applying a sequence of plane rotations to a matrix).

4.3   Block Algorithms

In some algorithms (e.g., computing the eigenvalues of a symmetric tridiagonal matrix), use of the BLAS is not feasible. In the majority of algorithms, however, there is scope for using the Level 2 and Level 3 BLAS; indeed, a considerable amount of experience has already been accumulated [14,22,15]. To exploit the Level 3 BLAS, one usually must express the algorithm at the top level in terms of operations on submatrices (the so-called "blocks") as compared to vector- or scalar-oriented operations. Block algorithms for solving equation systems are discussed in [8,9,19,22,25,26]; block algorithms for orthogonal factorizations can be found in [1,2,7,5,6,21,25,29,30,33,34,35].

As an example of a block algorithm, we consider an algorithm for computing the QR factorization

$$A = QR$$

of a dense matrix. Here an $m \times n$ (w.l.o.g. $m \geq n$) matrix A is decomposed into an orthogonal $m \times m$ matrix Q and an upper triangular $m \times n$ matrix R. This decomposition is one of the basic tools of numerical linear algebra; for applications, see [28]. The traditional algorithm for computing the QR factorization [28, p.148] employs a sequence of Householder reductions

$$H = I - 2uu^T, \quad \|u\|_2 = 1. \tag{4.1}$$

Application of H to a given matrix A involves a matrix-vector multiplication $z \leftarrow A^T u$ and a rank-one update $A \leftarrow A - 2uz^T$. Each of these operations requires $O(n^2)$ floating-point operations and uses $O(n^2)$ data.

To arrive at a block formulation of the Householder QR algorithm, we must be able to express a series of Householder reductions in a convenient closed form. Bischof and Van Loan [7] expressed the product

$$Q = H_1 \cdots H_b$$

of a series of $m \times m$ Householder matrices (4.1) in the so-called *WY representation*

$$Q = I + WY^T \tag{4.2}$$

where W and Y are $m \times b$ matrices. Schreiber and Van Loan [35] and independently Du Croz [23] refined this representation by expressing $W = YT$ where T is a $b \times b$ upper triangular matrix. Schreiber and Van Loan called the resulting representation

$$Q = I + YTY^T \tag{4.3}$$

the *compact WY representation* since it requires only about half as much storage as the original WY representation (4.2) in the typical case where $m \gg b$. Compared to the traditional Householder algorithm, the accumulation of T requires $O(mb^2)$ extra flops and $b^2/2$ extra words for storage. Since typically $m \gg b$, this is a low-order term in the overall algorithmic complexity. The advantage of the compact WY representation is that the computation of $A \leftarrow Q^T A$ now involves two matrix-matrix multiplications

$$Z \leftarrow A^T YT \tag{4.4}$$

and a rank-b update

$$A \leftarrow A + YZ^T \tag{4.5}$$

instead of a series of b matrix-vector multiplications and rank-one updates.

We can now express the block Householder QR algorithm in terms of the primitives *gencwy* (generate compact WY factor) and *appcwy* (apply compact WY factor):

$$[Y, T] \leftarrow gencwy(A)$$

returns the compact WY factors T and Y such that

$$A = (I + YTY^T)\begin{pmatrix} R \\ 0 \end{pmatrix}.$$

The primitive *gencwy* first computes the QR factorization of A using the traditional Householder QR algorithm and then accumulates T. Next,

$$A \leftarrow appcwy(Y, T, A)$$

performs the updates (4.4) and (4.5). Algorithm 4.1 shows the block Householder algorithm using the compact WY representation. Here A is partitioned as an $M \times N$ block matrix, and for simplicity we assume that all blocks are of the same size $b_m \times b_n$, so $m = Mb_m$ and $n = Nb_n$. We use the notation $A(i, j)$ to refer to block entry (i, j) and $A(i : j, k : l)$ to refer to the submatrix of A consisting of block row entries i to j and block column entries k to l.

Algorithm 4.1 The Bock Householder QR Factorization Algorithm.

> **for** $i = 1$ **to** N **do**
> $[Y, T] \leftarrow gencwy\ (A(i\!:\!M, i))$
> $A(i\!:\!M, i\!:\!N) \leftarrow appcwy\ (Y, T, A(i\!:\!M, i\!:\!N))$
> **end for**

This algorithm illustrates some important features of the block algorithm. For one, the block algorithm may require more floating point operations than its unblocked counterpart. We invest more work in accumulating a block transformation, but this is more than made up for by the application of the transformation, which will run at close to optimum speed since it is not slowed down by excessive data movement overhead. This reasoning is true up to the point where adding more columns to a block transformation will not result in a faster update.

This relates to the subtle issue of partitioning a given matrix into blocks. The block partitioning resulting in the fastest execution of the code (the "optimal" block partitioning) is problem-dependent (we can use larger blocks for larger matrices), but it also depends on the architecture of a given machine. Furthermore, on multiprocessor machines, possibly conflicting issues of individual processor performance and overall load balancing must be reconciled. A discussion of these issues and a suggestion for a methodology to overcome this problem can be found in [4]. Determining optimal, or near optimal, block sizes for different environments is a major research topic for the LAPACK project.

Algorithm 4.1 constructs a block transformation and then immediately applies it to the remaining sub-matrix. We will call this a *"right-looking algorithm"*. Notice that at each of the N stages we are updating a submatrix of size $(m - (i - 1)b_m + 1) \times (n - (i - 1)b_n + 1)$. We can further reduce the amount of data movement as indicated in Algorithm 4.2.

Algorithm 4.2 Left-Looking Block Householder QR Factorization Algorithm.

> **for** $i = 1$ **to** N **do**
> > **for** $j = 1$ **to** $i - 1$ **do**
> > > $A(j:M,i) \leftarrow appcwy\ (Y_j, T_j, A(j:M,i))$
> >
> > **end for**
> > $[Y_i, T_i] \leftarrow gencwy\ (A(i:M,i))$
>
> **end for**

At each stage of this algorithm we are only modifying a $m \times b_n$ matrix. We will call this algorithm the *"left-looking algorithm"*. Compared to Algorithm 4.1 this algorithm requires as many reads, but the number of writes is substantially reduced. This is particularily important in shared memory multiprocessors where cache consistency is guaranteed by the use of "write-through" caches [31,37]. On those architectures read accesses to cached data can be satisfied in one cycle, but write accesses are immediately flushed to memory. As a result, write accesses are much slower than read accesses.

4.4   Target Machines

The library will be designed primarily to perform efficiently on machines with a modest number of processors (say, 1-100), each having a powerful vector-processing capability. These machines include all of the most powerful computers currently available and in use for general-purpose scientific computing: CRAY-2, CRAY X-MP, CRAY Y-MP, CYBER 205, ETA-10, Fujitsu/Amdahl VP, IBM 3090/VF, NEC SX, Alliant FX/80, Convex C-1, Convex C-2, Scientific Computer Systems SCS-40, Ardent, Stellar, Sequent Symmetry, Encore Multimax, and BBN Butterfly. We hope that the library will also perform well on a wider class of parallel machines, including the Ametek, IBM RP3, Intel iPSC, and NCUBE. On conventional serial machines, the performance of the library is expected to be at least as good as that of the current LINPACK and EISPACK codes. Thus the library will be suitable across the whole range of machines from personal computers to supercomputers to experimental architectures.

We do not claim that the strategy of using Level 2 or Level 3 BLAS will necessarily attain optimal performance on all these machines; indeed, some algorithms can be structured in several different ways, all calling Level 3 BLAS, but with different performance characteristics. In such cases we will choose the structure that provides the best "average" performance over the range of target machines. Currently we are limiting machine-dependent optimizations to the BLAS to retain portability across architectures. We encourage vendors to provide implementations of the BLAS kernels that are optimized for their particular architecture. While users are free to develop their own versions of the LAPACK codes, we believe that the possible performance gain will be limited on the more conventional architectures.

On the more experimental architectures (in particular, distributed-memory machines), the restriction of optimization to the BLAS might be too limiting. In particular, it might be advantageous to introduce parallelism at the top-level of the algorithm instead of inside the BLAS. To aid users in experimenting on their particular architecture, the LAPACK codes have been carefully designed in a modular fashion and with the objective of minimizing data movement. Since data movement is the key issue in distributed-memory as well as shared-memory machines, the LAPACK codes should be easily "tunable" to more experimental architectures.

Several of the algorithms we intend to implement [16] will require more than loop-based parallelism. These algorithms will rely upon the simplified SCHEDULE mechanism [17] to invoke parallelism. These ideas might also be used to express top-level parallelism in a portable fashion. We are also closely following the activities of the Parallel Computing Forum [24] which has been formed by computer vendors, software developers, national laboratories, and universities to exchange technical information and to document agreements on constructs for programming parallel applications for shared-memory multiprocessors.

4.5 Programming Language and Style

The software will be developed in standard Fortran 77, with extensions to the standard only where necessary. Single- and double-precision versions will be prepared; conversion between different precisions will be performed automatically by software tools.

Routines for complex matrices will use the COMPLEX data type (like LINPACK, but unlike EISPACK); hence the availability of a double-precision complex (COMPLEX*16 or DOUBLE COMPLEX) data type will be assumed as an extension to Fortran 77. Routines for real and complex matrices will be written to maintain a close correspondence between the two and to permit automatic transformation, as far as possible; however, in some algorithms (e.g., unsymmetric eigenvalue problems) the correspondence will necessarily be weaker.

We realize that Fortran 8X is likely to have a number of features that would improve the design and coding of the library. In particular, its built-in array features would replace some of the BLAS kernels and result in cleaner and easier-to-read code. Another useful feature is the dynamic allocation of workspace. Almost all block routines need work space, with the optimal amount of storage determined depending on the problem parameters at run-time. Currently, the user will pass a work array in the argument list in the hope that it will be big enough; if it is not, the block size used may be less than optimal. In Fortran 8X, work space could be allocated dynamically at run-time in a fashion that is transparent to the user. This would significantly shorten calling sequences and avoid some common programming mistakes resulting from passing too little work space.

We are also planning to provide a C version of LAPACK which will be produced automatically from the Fortran version. However, since we are to begin development and testing of the library now on our range of target machines, Fortran 77 seems the reasonable choice.

4.6   User Interface and Documentation

The user interface to the routines will be similar to that of LINPACK. Routine names and arguments will follow a systematic scheme (see [3]).

We intend to provide a set of top-level driver routines that can solve a complete problem (e.g., compute the QR factorization of a given matrix). Each routine will call a number of lower level routines which perform the parts of the solution (e.g., generation of Householder vectors, generation of block transformations, application of block transformations).

As an example, the calling tree for this particular application is shown in Figure 4.1. SLARFG generates a Householder vector, SLARF applies a Householder vector, SGEQR2 computes an unblocked QR decomposition, SLARFT accumulates a block Householder matrix, SLARFG applies a block Householder matrix, and SGEQRF computes a QR factorization of a real matrix in single precision using the block algorithm.

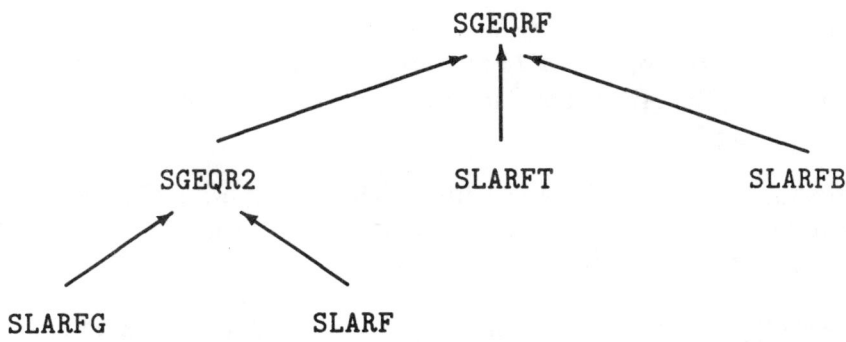

Figure 4.1: Calling Tree for SGEQRF.

Both levels of routines will be accessible to users and will be documented in a guide, similar in many respects to the LINPACK Users' Guide [11]. As a result, LAPACK will provide easy-to-use interfaces for the user who simply wants to solve a particular problem with minimal effort, but will also serve as a powerful toolbox for the more ambitious user interested in designing new algorithms.

References

[1] Bai, Z., and J. Demmel, *LAPACK Working Note #8: On a Block Implementation of Hessenberg Multishift QR Iteration*, Technical Report ANL–MCS–TM–127, Argonne National Laboratory, Mathematics and Computer Sciences Division, 1989.

[2] Berry, M., K. Gallivan, W. Harrod, W. Jalby, S. S. Lo, U. Meier, B. Philippe, and A. Sameh, "Parallel Algorithms on the Cedar System," *Proceedings of CONPAR 86*, W. Häandler (ed.), Springer Verlag, New York, 25–39, 1986.

[3] Bischof, C., J. Demmel, J. Dongarra, J. D. Croz, A. Greenbaum, S. Hammarling, and D. Sorensen, *LAPACK Working Note #5: Provisional Contents*, Technical Report ANL–88–38, Argonne National Laboratory, Mathematics and Computer Sciences Division, September 1988.

[4] Bischof, C. H., *Adaptive Blocking*, Technical Report ANL/MCS–P39–1288, Argonne National Laboratory, Mathematics and Computer Sci-

ences Division, 1988.

[5] Bischof, C. H., *Computing the Singular Value Decomposition on a Distributed System of Vector Processors*, Technical Report 87–869, Cornell University, Department of Computer Science, 1987. To appear in Parallel Computing.

[6] Bischof, C. H., "A Pipelined Block QR Decomposition Algorithm," in *Parallel Processing for Scientific Computing*, G. Rodgrigue (ed.), SIAM, Philadelphia, 3–7, 1989.

[7] Bischof, C. H., and C. F. Van Loan, "The WY Representation for Products of Householder Matrices," *SIAM J. Sci. Stat. Comput.*, 8, s2–s13, 1987.

[8] Calahan, D., "Block-Oriented Local-Memory-Based Linear Equation Solution on the Cray-2: Uniprocessor Algorithms," in *Proceedings International Conference on Parallel Processing*, IEEE Computer Science Press, August 1986.

[9] Dave, A. K., and I. S. Duff, "Sparse Matrix Calculations on the CRAY-2," *Parallel Computing*, 5, 55–64, 1987.

[10] Demmel, J., J. D. Croz, S. Hammarling, and D. Sorensen, *Guidelines for the Design of Symmetric Eigenroutines, SVD and Iterative Refinement for Linear Systems*, Technical Report LAPACK working note #4, Argonne National Laboratory, Mathematics and Computer Sciences Division, February 1988.

[11] Dongarra, J. J., J. R. Bunch, C. B. Moler, and G. W. Stewart, *LINPACK Users' Guide*, SIAM Philadelphia, 1979.

[12] Dongarra, J., J. D. Croz, I. Duff, and S. Hammarling, *A Set of Level 3 Basic Linear Algebra Subprograms*, Technical Report MCS-P1-0888, Argonne National Laboratory, Mathematics and Computer Sciences Division, August 1988.

[13] Dongarra, J., and E. Grosse, "Distribution of Mathematical Software by Electronic Mail," *Comm. ACM*, 30, 5, 403–407, 1987.

[14] Dongarra, J., S. Hammarling, and L. Kaufman, "Squeezing the Most Out of Eigenvalue Solvers on High-Performance Computers," *Linear Algebra and Its Applications*, 77, 113–136, 1986.

[15] Dongarra, J., A. Sameh, and D. Sorensen, "Implementation of Some Concurrent Algorithms for Matrix Factorization," *Parallel Computing*, 3, 1, 25–34, 1986.

[16] Dongarra, J., and D. Sorensen, "A Fully Parallel Algorithm for the Symmetric Eigenvalue Problem," *SIAM J. Sci. Stat. Comput.*, 8, 2, 139–154, 1987.

[17] Dongarra, J., and D. Sorensen, "A Portable Environment for Developing Parallel Programs," *Parallel Computing*, 5, 1&2, 175–186, 1987.

[18] Dongarra, J. J., J. D. Croz, S. Hammarling, and R. J. Hanson, "An Extended Set of Fortran Basic Linear Algebra Subprograms," *ACM Trans. Math. Software*, 14, 1, 1–17, 1988.

[19] Dongarra, J. J., J. D. Croz, I. Duff, and S. Hammarling, *A Set of Level 3 Basic Linear Algebra Subprograms*, Technical Report ANL–MCS–TM88 (Revision 1), Argonne National Laboratory, Mathematics and Computer Sciences Division, May 1988.

[20] Dongarra, J. J., and I. S. Duff, *Advanced Computer Architectures*, Technical Report ANL–MCS–TM57, Argonne National Laboratory, Mathematics and Computer Sciences Division, Revision 1, 1987.

[21] Dongarra, J. J., S. J. Hammarling, and D. C. Sorensen, *Block Reduction of Matrices to Condensed Form for Eigenvalue Computations*, Technical Report ANL–MCS–TM99, Argonne National Laboratory, Mathematics and Computer Sciences Division, September 1987.

[22] Dongarra, J. J., and D. C. Sorensen, "Linear Algebra on High-Performance Computers, in *High-Performance Computers 85*, Udo Schendel, (ed.), North-Holland, Amsterdam, 3–32, 1986.

[23] Croz, J. D., Private Communication, 1987.

[24] The Parallel Computing Forum, *PCF Fortran: Language Definition*, Kuck and Associates, Champaign, IL 61820, 1988.

[25] Gallivan, K., W. Jalby, U. Meier, and A. Sameh, "The Impact of Hierarchical Memory Systems on Linear Algebra Algorithm Design," *SIAM J. Sci. Stat. Comput.*, 8, 6, 1079–1084, 1987.

[26] Gallivan, K., W. Jalby, U. Meier, and A. Sameh, *The Impact of Hierarchical Memory Systems on Linear Algebra Algorithm Design*, Technical Report 625, University of Illinois at Urbana-Champaign, Center for Supercomputing Research and Development, September 1987.

[27] Garbow, B., J. Boyle, J. Dongarra, and C. Moler, *Matrix Eigensystem Routines – EISPACK Guide Extension*, Volume 51 of *Lecture Notes in Computer Science*, Springer Verlag, New York, 1977.

[28] Golub, G. H., and C. F. Van Loan, *Matrix Computations*, The Johns Hopkins University Press, 1983.

[29] Harrod, W., *Solving Linear Least Squares Problems on an Alliant FX/8*, Technical Report, University of Illinois at Urbana-Champaign, Center for Supercomputing Research and Development, 1986.

[30] Higham, N., and R. Schreiber, *Fast Polar Decomposition of an Arbitrary Matrix*, Technical Report, Cornell University, Department of Computer Science, 1988.

[31] Hwang, K., and F. A. Briggs, *Computer Architecture and Parallel Processing*, McGraw-Hill, New York, 1984.

[32] Lawson, C. L., R. J. Hanson, R. J. Kincaid, and F. T. Krogh, "Basic Linear Algebra Subprograms for Fortran Usage," *ACM Trans. Math. Software*, 5, 3, 308–323, 1979.

[33] Schreiber, R., *Block Algorithms for Parallel Machines*, Technical Report 87–5, Rensellaer Polytechnic Institute, Department of Computer Science, 1987.

[34] Schreiber, R., and B. Parlett, *Block Reflectors: Computation and Applications*, Technical Report, Rensellaer Polytechnic Institute, Department of Computer Science, 1987.

[35] Schreiber, R., and C. Van Loan, *A Storage Efficient WY Representation for Products of Householder Transformations*, Technical Report CS–87–864, Cornell University, Department of Computer Science, 1987.

[36] Smith, B., J. Boyle, J. Dongarra, B. Garbow, Y. Ikebe, V. Klema, and C. B. Moler, *Matrix Eigensystem Routines – EISPACK Guide*, Springer Verlag, New York, second edition, 1976.

[37] Stenström, P., "Reducing Contention in Shared-Memory Multiprocessors," *IEEE Computer*, 21, 11, 26–37, 1988.

[38] Stone, H., *High-Performance Computer Architecture*, Addison-Wesley, Reading, Massachussetts, 1987.

Chapter 5

Parallel Sub-Domain and Element-by-Element Techniques

*G. F. Carey**

5.1 Scientific Computing

The recent and continuing revolution in microelectronics has profoundly changed the way we think about scientific computing. The power of microprocessors has increased dramatically while their relative cost has been reduced. This has led to a rapid expansion in personal computers and workstations. Some scientific workstations can now offer faster computer speed than many mainframe systems of 5 years ago at a fraction of the cost. Vector "boards" are being incorporated into these systems to further enhance performance. For problems of moderate size these advanced scientific workstations provide a very suitable computing environment. At the same time, microelectronic miniaturization advances have allowed us to push high speed supercomputing to new levels. This is essential if large scale complex three-dimensional simulations are to be addressed in reasonable computation time. Examples of the latter applications include weather forecasting, "war games", complex non-linear dynamics and chaos, oil reservoir simulation, manufacturing processes, geological and environmental problems and economic modelling to name only a few important topics. Full scale simulations of problems in these areas may

---

*The University of Texas, Austin, TX.

require days of computer calculation on even standard mainframe systems. In the case of short-range weather forcasting and war strategy logistics, this delay will, obviously, defeat the purpose of the computation. In other instances the increase in flow time for a computation will restrict the number of analyses possible (e.g., design modifications and preliminary design studies) and significantly extend the project time or limit the design. For best productivity, it could be argued that the analyst should not have to wait more than one or two hours for a large-scale simulation. Finally, there are still many scientific problems such as those involving chaos and turbulence in 3-D flows that remain beyond the scope of reasonable present day computation.

Thus we see that there are two main forces at play — the inexpensive mass production of powerful microprocessors and the need to solve very large scale complex problems. Both of these forces are active in driving the technology. However, the scale of the microprocessor components is now at the submicron level and further dramatic reductions in size are not possible with existing technology. It will take a major technological breakthrough (perhaps involving quantum tunneling or Josephson junction devices and superconductivity) to continue to achieve orders of magnitude gain in single processor speedup in timeframes similar to those witnessed in the past decade. An alternative strategy is to seek increases in computational power by parallel processing. Since processors can be mass produced more efficiently, many processors can be configured in parallel to share the computational task. This has led to a variety of concurrent computing systems ranging from a few very sophisticated processors (coarse granularity) as in the CRAY X-MP series to many parallel simple processors (fine granularity) as in the CONNECTION machine. A second important distinction is the choice of a shared memory environment as in the CRAY, ETA, ALLIANT, and Balance systems or a distributed memory associated with individual processors such as the INTEL, NCUBE and CONNECTION hypercube configurations.

5.2   Methods and Algorithms

Clearly, the variety of available and emerging systems implies that there are important issues that must be addressed related to the choice of algorithms that will be suited to parallel computation. To a lesser extent the same issue arose when vector processors were introduced and efficient vectorization strategies became the focus of supercomputer algorithm research. However, for vector supercomputers, the development was more gradual and the issues less varied and complex. Now, suprisingly, the development of parallel processors has been so rapid that research on software and algorithms lags the hardware. This problem is being addressed as evident in the 1986 special issue of Communications in Applied Numerical Methods devoted to Algorithms for Supercomputers. Journals on parallel processing have recently been formed and the topic is the subject of special sessions at conferences and symposia.

As in the case of vector supercomputing, considerable effort is being invested in the development of compilers that will carry out parallelization automatically. In this way existing software can be conveniently taken into the concurrent processing regime. Additional compiler directive instructions permit the analyst to take advantage of parallel processing opportunities and restructure the program to achieve better speed-up. It is particularly important that the algorithms for the standard library routines and intrinsic functions be efficient. Similarly, frequently used numerical analysis libraries such as "LINPACK" must be transformed to a highly efficient concurrent form for the parallel processing environment. Such transformations may not be straightforward. Von Neumann's remark in his 1954 report on the Advanced Institute Machine seems particularly appropriate in this context: "it will change our entire way of viewing what is elegant or clumsy." In the same sense, the advent of parallel processing has had a profound effect on our assessment of the efficiency of algorithms and the ease with which they may be parallelized. It also leads to the possibility of introducing asynchronous schemes which admit an element of randomness to the order of computations and to the numerical results. Of course, the real challenge is to map algorithms efficiently to a parallel computing environment. Rather than consider a specific parallel computer configuration, one can also elect to study methods and algorithms, examining the opportunities for parallel computing and configure a parallel architecture accordingly.

There are several levels at which one can approach parallel processing in mathematical modelling. For instance, physical, chemical and other processes that evolve simultaneously could suggest a parallel decomposition of the problem. Previously we have referred to this as *inherent* parallelism (Carey [6]). On the other hand, algorithm considerations may influence the decision. To set this in perspective, it is useful to review the main steps in computer simulation of a physical or design process:

1. First the main physical processes are identified;

2. Then appropriate conservation laws and constitutive relations are introduced that lead to a mathematical statement of the problem (for example, as a PDE system);

3. A discretization procedure leads to an approximate formulation (as, perhaps, an algebraic system or system of ordinary differential equations);

4. Numerical techniques are introduced and efficient algorithms constructed to implement these procedures and solve the discrete system;

5. Finally the results are post-processed to calculate other quantities of interest (fluxes, integrals) from the numerical solution.

As an illustrative example related to inherent parallelism in a physical model, let us consider fluid flow and transport. The momentum equations for fluid flow may be coupled to heat and mass transport equations, turbulence transport equations, and perhaps electromagnetic field equations. A natural iterative decoupling of the equations suggests a parallel computing system of coarse granularity (few processors) where the discretized momentum equations are solved on one processor and discretized transport equations on the other processors. Different numerical grids could be used on respective processors to better capture relevant details of the respective solution variables such as boundary and interior layers or singularities. The problem of efficiently balancing the workload between processors, however, would be more difficult in such a scheme due to the differences in grid size. Alternatively, all processors could be employed in the solution algorithm for each discretized equation using, for instance, domain decomposition ideas.

5.3   Domain Decomposition Concepts

At the discretization level, there are several possibilities for exploiting parallelism. The most obvious is to partition the problem domain to subdomains and distribute subdomains to different processors. A simple example of this procedure can be drawn from numerical grid generation. Here we wish to generate a numerical grid discretizing a general domain in two or three dimensions. The domain may be geometrically complex and this makes smooth global grid generation by, for instance, conformal mapping from the unit square or cube reference grids difficult. Instead the physical domain can be partitioned to a union of disjoint subdomains with grid points specified on the subdomain boundaries. Then algebraic or PDE grid generation algorithms can be applied in parallel and completely independently on the subdomains. Of course, the number of subdomains, size of subdomain grids, and number of processors will impact the load balance characteristics and performance of this simple domain decomposition strategy. Local grid smoothing can also be carried out in parallel and then the interface grid points adjusted between subdomains. This last step will require communication between processors associated with the respective subdomains.

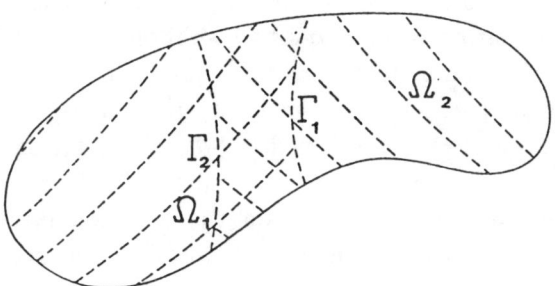

Figure 5.1: Schwarz Decomposition.

Related domain decomposition ideas can be also utilized in solving a discretized boundary-value problem, and this has become a topic of recent research interest (Glowinski *et al.* [12,13]). In fact this idea dates back to the work of Schwarz in the 19th century. Schwarz [18] proposed solving the Poisson equation

$$\Delta u = f \quad \text{in } \Omega, \quad u = g \quad \text{on } \partial\Omega \tag{5.1}$$

by decomposing the domain to two overlapping subdomains Ω_1 and Ω_2 as indicated in Figure 1, and then solving the subdomain problems. During this process the interior solution on each individual subdomain is used to update the overlap boundary data on Γ_1 and Γ_2 within the adjacent subdomains. For example, in Figure 5.1, a starting guess α may be specified for the solution on line Γ_1 inside Ω_2. The first subdomain problem becomes

$$\Delta u_1 = f \quad \text{in } \Omega_1$$

with

$$u_1 = g \quad \text{on } \partial\Omega_1 \cap \partial\Omega$$

and

$$u_1 = \alpha \quad \text{on } \Gamma_1. \tag{5.2}$$

The solution u_1^0 in Ω_1 provides the boundary data β on Γ_2 for solution of a similar problem on Ω_2. This, in turn, yields an improved iterate on Γ_1, and the procedure can be repeated. By means of the maximum principle, convergence of this recursive iteration can be proven. The procedure can be easily adapted to parallel form by providing initial starting iterates on Γ_1 and Γ_2 simultaneously and solving concurrently. The idea can also be directly generalized to treat multiple overlapping domains, or the process can be applied repeatedly to each of the two initial subdomains by using repeated subdomain decomposition from 2 to 4 to 8 subdomains, and so on.

Other forms of domain decomposition can be constructed and may be preferable. For example, let there be no overlap in Figure 5.1, so that Γ_1 and Γ_2 coincide. The physics implies that both the solution and normal flux $\sigma \cdot n = \frac{\partial u}{\partial n}$ should be continous across Γ. Accordingly, let the same initial iterate be specified as Dirichlet data on the common interface Γ and the distinct problems solved concurrently in Ω_1 and Ω_2. The respective solutions can be post-processed to determine accurately the associated normal fluxes at Γ. For the exact solution the normal flux jump across Γ should be zero, and this condition can be used to adjust the interface fluxes. Separate problems with Neumann boundary conditions on Γ can then be solved in the second step. This determines two distinct new values of u on Γ, which may be appropriately averaged to satisfy continuity. The two-step procedure can then be repeated.

This idea can be applied to mixed methods for the solution pair (u, σ) satisfying the first order system

$$\nabla \cdot \sigma = f, \quad \sigma - \nabla u = 0, \quad \text{in } \Omega \tag{5.3}$$

using, for instance, a mixed finite element method. Various other generalizations of these concepts can be developed including, perhaps, iterative solutions on the subdomains. If one considers the discretized system for the entire domain, then the approach can be interpreted as a form of block (or subdomain) iteration. In this way we can recast this class of methods within the structure of iterative and block iterative schemes. We take up some related issues later for element-based block iterative schemes.

A different class of subdomain approaches has been extensively applied in structural analysis in the form of substructure methods. Large scale structural discretizations that could not be stored in core are partitioned to separate substructures Ω_i and pre-elimination of interior solution variables carried out to construct reduced substructure system contributions (subsystem condensation). The resulting substructure matrices can be thought of as super-element matrices. These reduced substructure matrices can then be assembled to a global system matrix and solved in-core. For example, let $Au = b + \sigma$ denote the equations corresponding to substructure i. Here b are nodal forces and σ are nodal tractions at the substructure boundary. Partitioning to block form with u_r and u_c, the retained and condensed variables, respectively, and writing f for the right side, we get

$$A_{rr} u_r + A_{rc} u_c = f_r$$

$$A_{cr} u_r + A_{cc} u_c = f_c \tag{5.4}$$

From the latter equation

$$u_c = A_{cc}^{-1}\left(f_c - A_{cr}u_r\right) \tag{5.5}$$

Substituting in the first equation and collecting terms we get

$$\left(A_{rr} - A_{rc}A_{cc}^{-1}A_{cr}\right)u_r = f_r - A_{rc}A_{cc}^{-1}f_c \tag{5.6}$$

as the reduced system. That is, the matrix $\tilde{A} = A_{rr} - A_{rc}A_{cc}^{-1}A_{cr}$ is the reduced coefficient matrix for block Ω_i. Note that on assembly of the subdomain contributions the normal flux jump condition $[\![\sigma_n]\!] = 0$ on the subdomain interfaces will be enforced. The element matrix calculations and subdomain condensation can be locally vectorized and carried out independently in parallel over the substructures. Assembly of the global system involves communication between processors in a distributed memory system, but can be accumulated directly in a shared memory configuration. Solution of the global reduced system can then use a parallel iteration or elimination method. Finally, the substructure solutions can be post-processed in parallel to yield the nodal values at the interior condensation points. Again, substructuring can be interpreted at the algebraic level in a more abstract sense by block partitioning the assembled large global system to separate entries corresponding to nodal unknowns that are to be retained and condensed, respectively.

5.4   Iteration and Block Methods

In the preceeding discussion, we observed that certain subdomain strategies were equivalent to block iterative methods, in a partitioning of the global system. There are alternative abstract forms of partitioning that can also be used to enhance vectorization and parallelization. For example, the classical Jacobi iteration for solution of a linear system updates the solution value of each component using the old iterate values of all other components of the solution vector. Equivalently, for linear system

$$Au = b, \tag{5.7}$$

we have at iterate $k + 1$ and component i

$$u_i^{(k+1)} = u_i^{(k)} + \frac{r_i^{(k)}}{A_{ii}} \tag{5.8}$$

where $r^{(k)} = b - Au^{(k)}$ is the residual. In this residual-based iterative method, each component may be updated independently, and therefore in parallel. In the case of a sparse system resulting from a finite difference or finite element discretization, the calculation involves only old nodal values at grid points immediately adjacent to the node i in question. We can interpret the diagonal blocks as simple diagonal entries A_{ii} and, if desired, associate a single processor with each node of the partition. A similar argument applies to explicit methods for transient problems.

Now it is well know that the Jacobi iteration is very slow, and the Gauss-Seidel iteration is faster on sequential scalar machines. In Gauss-Seidel iteration the most recently computed nodal values during the iterative sweep are used in updating the current value u_i, so (5.8) becomes

$$u^{(k+1)} = A^{-1} \left(b - A^L u^{(k+1)} - A^U u^{(k)} \right) \tag{5.9}$$

where $A^L D^{-1} A^U$ are the lower triangular, diagonal, and upper triangular parts of A. Note that the algorithm is now recursive and this feature limits the vectorization and concurrency. For example, natural ordering of grid points (left to right and bottom to top) in a square domain will not permit efficient vector or parallel processing. Ordering nodes diagonally from "northwest to southeast" beginning at the bottom left corner will circumvent this problem to some degree. Vector lengths are then proportional to the number of nodes per diagonal and for parallel calculations with shared memory, the parallel synchronization lags by one diagonal. This can be effective for such simple regions but for general domains it is of limited value. One can exploit the concurrency of the Jacobi method and avoid the recursive difficulties by using a red-black iterative scheme: In a natural ordering the nodes are alternately "colored" red and black as in a checkerboard. Then the block partitioning is introduced based on this coloring. Since any red node has four black neighbours and vice versa, we can iterate in parallel over the red nodes and the black nodes in a dual-pass procedure using successive over relaxation. This idea can be generalized further to include other colorings such as multicolor iterative methods (see Adams [1]).

More generally, an iterative algorithm for solving $Au = b$ involves a basic iterative method and an acceleration procedure. Such an iterative scheme can be written in the form

$$u^{(k+1)} = Gu^{(k)} + s \tag{5.10}$$

where $G = I - Q^{-1}A$ and $s = Q^{-1}b$ (e.g., see Hageman and Young [14]). The matrix G is termed the splitting matrix and the preconditioner Q is usually chosen so that it is easy to solve the auxiliary system $Qx = y$. For example, in the Jacobi iteration, $Q = D$ where D is the diagonal matrix $D = \text{diag}(A)$.

The preconditioned conjugate gradient method is well suited to symmetric problems and often can be applied successfully to slightly nonsymmetric problems. For general nonsymmetric problems other gradient-type iterative methods are preferable. The Lanczos method and particularly truncated forms of this scheme (biconjugate gradient method — BCG) are effective but are not guaranteed to converge. Restarting the BCG method as "breakdown" is approached has circumvented this problem in all practical applications we have computed. The conjugate gradient squared (CGS) method is frequently superior to BCG but is less robust. Other residual type gradient methods such as Orthomin, Orthores, and GMRES methods may be similarly applied.

Central to these gradient methods are repeated matrix-vector products. We see this, for example, in the BCG algorithm: Given starting iterate u^0, compute

$$u^{(k+1)} = u^{(k)} + \beta_k p^{(k)} \tag{5.11}$$

using

$$\delta^{(0)} = Q^{-1}(b - Au^0), \qquad \bar{\delta}^{(0)} = \bar{\delta}^{(0)} = p^{(0)} = \bar{p}^{(0)}$$

$$\delta^{(k+1)} = \delta^{(k)} - \beta_k Q^{-1} Ap^{(k)}, \qquad \bar{\delta}^{(k+1)} = \bar{\delta}^{(k)} - \beta_k (Q^{-1}A)^T \bar{p}^{(k)} \tag{5.12}$$

where

$$p^{(k)} = \delta^{(k)} + \alpha_k p^{(k-1)}, \qquad \bar{p}^{(k)} = \bar{\delta}^{(k)} + \alpha_k \bar{p}^{(k-1)} \tag{5.13}$$

and

$$\beta_k = \frac{\delta^{(k)T}\bar{\delta}^{(k)}}{\left(\bar{p}^{(k)}\right)^T Q^{-1} Ap^{(k)}}, \qquad \alpha_k = \frac{\delta^{(k)T}\bar{\delta}^{(k)}}{\delta^{(k-1)T}\bar{\delta}^{(k-1)}} \tag{5.14}$$

with Q the preconditioning matrix for A. Note that in this form the algorithm involves repeated matrix vector products of the form $Av = w$. Efficient vectorization requires that these matrix-vector products be calculated with long vectors. Similarly, good concurrency and load balance is needed on parallel processors for this computationally intensive part of the algorithm. For vector-parallel systems, both goals are to be met.

Let us consider the matrix-vector product $w = Av$ written as

$$w = \sum_{j=1}^{N} v_j c_j \tag{5.15}$$

where c_j are the columns of A and v_j are the components of v. This can be easily vectorized with vector lengths of size N and is well suited to dense matrices A such as those obtained in boundary element methods and global expansion (full spectral) methods. If A is sparse as in finite difference or standard (low-degree) finite element methods, then a sparse matrix format is employed and vector lengths are not long. The approach (5.15) can also be used for parallelization over the columns with subsequent cyclic accumulation of vectors (addition of adjacent pairs of vectors, then resulting pairs, and so on). The number of vectors is halved at each step in the cyclic reduction so that at some point the number of vectors is less than the number of processors and load balance deteriorates.

If the gradient schemes are rephrased in a block iterative context, along the lines suggested previously, then the resulting algorithm requires block matrix-vector products $A_i v_i = w_i$ for each block i. In the previous subdomain setting, the block submatrix A_i would be associated with the discretization of subdomain Ω_i. For example, the elements constituting domain Ω_i may be generated and element matrices assembled on Ω_i to form A_i. The block matrix vector products could be vectorized for each subdomain in parallel according to (5.15) with subdomain contributions then accumulated to obtain the global vector desired. For load balancing, the submatrices A_i should be of comparable size, and the number of blocks an integer multiple of the number of processors.

Evidently, it is not essential that the element calculations be made in continuous blocks and, in fact, one can consider the matrix-vector product for an abstract domain decomposition in which each block consists of an arbitrary set of elements in the discretization of the domain. This arrangement enhances load balancing between processors since the block size is essentially constant and the number of blocks arbitrary. On the other hand this flexibility will lead to other complications and incur an additional computational cost proportional to the increase in the boundary measure of the subdomains as we explain later. As a compromise, one can develop an additional algorithm that employs contiguous blocks but with irregular interfaces chosen to improve load balance.

A further abstraction of the schemes may be deduced from the data structure that arises naturally from the finite element method, and this leads to the general concept of element-by-element parallel and vector-parallel methods (e.g., see Carey *et al.* [7]). To better understand this procedure let us first recall the basic steps in the standard finite element method: (1) The main feature is a loop over the elements in which the element matrix contributions A_e, b_e are computed based on a local element numbering system. (2) The element local-node to global-node relationship identifies the approximate row and column locations in the global matrix, and entries of A_e, b_e can be added directly into these locations. In practice this would also utilize a packed global storage format (such as a banded, envelope, or some similar format). Essential boundary conditions can be enforced at the element or system level.

(3) The final system can then be solved using elimination or iteration. Here we focus upon iterative solution using conjugate gradient type methods in an element-by-element block strategy.

Now the element calculations in step (1) are completely independent irrespective of element order and hence can be computed concurrently with near-perfect load balance over the processors. (Here, we assume that the problem size is a reasonable multiple of the number of processors.) The assembly in step (2) assumes that we really want to form the global system explicitly. Recalling the previous BCG algorithm in 5.11 to 5.15, we see that the global system enters principally in the matrix-vector product calculations. Rather than form global entities A, b and subsequently partition, we can use the element partitioning directly and sidestep the global assembly. To clarify this point, let us write $\hat{A}_e$ and $\hat{b}_e$ for the element contributions expanded to system size. That is, $\hat{A}_e = BA_e$ and $\hat{b}_e = Bb_e$, where B is the local to global Boolean map, are of size $N \times N$ and $N \times 1$ where N is the number of unknowns and the only non-zeros in $\hat{A}_e$ are those arising from A_e. Then the global assembly is formally equivalent to $A = \sum_e \hat{A}_e$ and the matrix vector product $w = Av$ can be recast as

$$w = Av = \left(\sum_{e=1}^{E} \hat{A}_e \right) v \qquad (5.16)$$

Further, let $\hat{v}_e$ contain the assembled entries associated with the nodes of element e and having all other entries zero. (That is $\hat{v}_e = xv$ for extraction operator x is an element "extraction" or "gather" from global vector v). Then in (5.16), since $\hat{A}_e$ is zero except in those columns associated with nodes of e, and recalling (5.5),

$$w = \sum_{e=1}^{E} \left(\hat{A}_e \hat{v}_e \right) = \sum_{e=1}^{E} \hat{w}_e \qquad (5.17)$$

The result $\hat{w}_e$ of the product $\hat{A}_e \hat{v}_e$ in (5.17) is a global vector that has nonzeros only in those components corresponding to the nodes of element e. This implies that we can directly compute the local contributions

$$w_e = A_e v_e , \qquad e = 1, 2, \ldots, E \qquad (5.18)$$

in parallel and then accumulate the results to global vector w by adding entries of w_e directly into the appropriate positions in w as defined by the global node numbering. For structured quadrilateral grids, the accumulation of local contributions w_e to a global vector can be easily parallelized by appropriately ordering the local node numbers. The local node numbers of a given global node should not be the same in different adjacent elements. More complex schemes are required for unstructured grids and efficiency is degraded.

The element matrix vector products in (5.18) can, of course, be vectorized as before. This is efficient if the vectors are sufficiently long. However, short vectors are usually obtained for standard finite elements of low degree. Only for p-methods which employ elements of very high degree will straightforward vectorization of the matrix product generally be useful.

There is a better strategy that can be easily applied to achieve good vectorization. Rather than loop over the elements and compute the product in (5.18) as the vectorized inner loop, we may reverse the loop order. That is, if the outer loop is over the local nodes and the inner loop is vectorized, this implies that vector length will be equal to the number of elements, which in practice is large. For this scheme the element matrices are stored in a 3-D array A_{eij} and the element vectors in a 2-D array v_{ei}. In the vectorized innermost loop, we compute for given i, j the pairwise products for each j: $v_{ei}A_{eij}$ with $e = 1, 2, \ldots, E$ and accumulate the resulting vectors for each i, j value in the outer loops. Parallelization now can be instrumented over the outer i, j loops. for standard elements of low degree, concurrency will be limited by the element matrix size. Bilinear elements for scalar potential problems in 2D generate 4×4 matrices and hence 16 vectors for calculation and accumulation. Biquadratic elements (9×9) have 81 entries and finer granularity is permissible. It is also possible to break the innermost loop over subdomains. Vector length is thereby reduced, but parallelism over the blocks can also be introduced. Note that, even if the elements assigned to a given processor are randomly ordered (to form an abstract block partition), this scheme will generate long vectors for that processer and hence be amenable to efficient vectorization.

5.5   Adaptive Grid Refinement

The procedures introduced here can also be applied to considerable advantage in conjunction with adaptive mesh refinement. In an adaptive finite element method, error indicators are computed from the approximate solution on a specified grid, and individual elements having large error indicators are subdivided. At the interfaces between refined and unrefined elements, constraints are introduced to ensure continuity of the approximation (conformity). This leads in practice to quite unstructured grids. Moreover, as new nodes are being introduced at each refinement step, the problem size increases. Since the bandwidth and envelope of the global system is very sensitive to global node numbering, elimination solvers fair poorly. Nodal renumbering schemes must then be introduced to optimize the band or envelope. Iterative methods are not limited in this way. Equally important, the solution on the previous coarse grid provides an accurate starting iterate for iterative system solution on the next grid. Finally the automated nesting of grids generated in this manner can be exploited by multigrid methods to accelerate fine grid convergence.

Our adaptive refinement scheme is an extension of the approach of Bank and Sherman [3]. An initial mesh of "macrofather" quadrilateral elements is prescribed and finer meshes generated by local refinement. At any step in the refinement process a given father element is subdivided to four sub-quadrilaterals ("sons"), to define successive levels in a tree data-structure. Locally, the nodes are numbered counterclockwise in a specific way. This implies that any regular node will have different local node numbers for the elements that share this node. Collecting element vector contributions w_e for each local node $i = 1, 2, 3, 4$ in (5.18), we can define four global vectors c_i of length equal to the number of nodes. The procedure for constructing c_i and their summation vectorizes. Nodes on the interior of the interface side between a refined and an unrefined element are termed irregular or constrained nodes and are an exception to the local node number rule above. The presence of these constrained nodes will lead to a slight degradation in the performance.

Since the concurrency in the element-by-element conjugate gradient type iterative schemes is over the elements, the parallel algorithm extends immediately to adaptively refined finite element grids. One can also vectorize over the elements to produce long vectors as indicated earlier. Achieving significant concurrency and long vectors simultaneously for massively parallel systems is more delicate but can be attained for very large problems.

5.6 Sample Results

Some representative results are briefly summarized to indicate the performance of the parallel element-by-element scheme on shared memory multi-processor systems such as the ALLIANT FX-8 and SEQUENT BALANCE. The problem in question corresponds to steady two-dimensional convection-diffusion in a square domain. A Galerkin finite element formulation is used with a discretization of 256 bilinear elements. Speedup-results are given in Figure 5.2 for parallel scalar element-by-element solution using biconjugate gradient iteration. The speedup was computed with reference to uniprocessor calculations for both the microtasked and unmicrotasked code. The difference in respective speedup is seen to be negligible. Slight degradation in speedup is observed at, for example, NP=7 processor performance since the load bal-

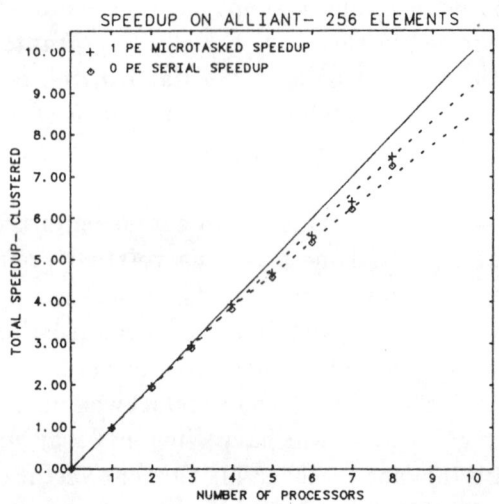

Figure 5.2: Speedup Performance.

ance is not then exact. At NP=8 there is perfect load balance in the EBE
procedure for 256 elements. Similar speedup results for this problem on the
SEQUENT BALANCE with 1024 elements and up to 20 processors together
with further details are given in Barragy and Carey [4] and Carey *et al.* [7].

The adaptive refinement scheme has been implemented for solution of
semiconductor device problems (Sharma and Carey [19]) using bilinear elements. Using adaptive refinement and iteration for the associated nonlinear
potential equation on the CRAY X-MP 24, we obtain the mesh refinement his-
tory and iteration count in Table 5.1. Scalar and vector uniprocessor results
are indicated. The adaptive refinement scheme has also been implemented
for scalar-parallel and vector-parallel calculations on the ALLIANT FX-8 in
some exploratory studies. Timing performance for the element-by-element
matrix-vector products per conjugate gradient (CG) iteration are summarized in Table 5.2. Results shown here are for the final mesh of 700 elements
(795 nodes).

In the EBE procedure described here the element matrix-vector products
are clearly independent and hence can be fully parallelized if the number of
elements is an integer multiple of the number of processors. This step of the
gradient iteration scheme therefore will parallelize well when the problem size
is large. As indicated earlier the elements can be distributed to processors in
random order without degradation. However, the element vectors obtained
from the EBE matrix-vector product are also to be accumulated to global
vectors. Since any global node will have several adjacent elements, this im-

Table 5.1: Timing for element-by-element scalar and vector calculations with adaptive refinement.

No. of Elements	No. of Iterations	Scalar Mode Exec. Time	Vector Mode Exec. Time
16	6	.00138	.00060
61	14	.00771	.00192
133	21	.02362	.00508
265	26	.05689	.01087
427	30	.10356	.01904

Table 5.2: Parallel scalar and vector performance for EBE matrix-vector product calculation on Alliant FX-8.

No. of Processors	Scalar EBE CG Mat.-Vec. Prod.	Vect. EBE CG Mat.-Vec. Prod.
1	$.676 \times 10^{-1}$	$.111 \times 10^{-1}$
2	$.343 \times 10^{-1}$	$.602 \times 10^{-2}$
4	$.176 \times 10^{-1}$	$.369 \times 10^{-2}$
8	$.938 \times 10^{-2}$	$.263 \times 10^{-2}$

plies that vector components from adjacent elements must be accumulated. If adjacent elements are, in fact, assigned to the same processor, then this poses no problem. On the other hand, if they are in different processors, the performance will be adversely affected. Hence, the previous "random" distribution will be less efficient overall, particularly for massively parallel distributed systems.

The communication overhead associated with assembly of neighbour contributions is directly proportional to the "active boundary measure" associated with the elements belonging to each processor. If the elements within each processor are contiguous, so that each processor has a simple block or subdomain, then the "active boundary" will be the interface boundary between processor blocks. This will lead to a reasonable scheme with modest communication overload. If further the subdomain interface measure is minimized by defining the blocks appropriately and with due consideration to load balance, then a very efficient scheme will result. For example, if a two dimensional rectangular domain is discretized to an $M \times N$ mesh with $M << N$, then the subdomain division can be easily made so that block interfaces are of size M (i.e., across the shorter section). Any interior block has $2(M+1)$ nodes on the two "active interfaces." The block size should be approximately constant (each containing, say, $Q \times M$ elements with $N/Q = P$ the number of processors available). Within each block the vectorized element-by-element calculations would be made as before, with vector accumulation from adjacent elements.

   Remarks: In fact, construction of the element matrices A_e is really not necessary (although it is convenient when one considers the structure of existing finite element computer programs). Recall that the matrix-vector product $Av = \omega$ corresponds to addition of vectors $\sum_{i=1}^{N} v_i c_i$ where c_i denote the columns of A. Applying this to the element product $A_e v_e = \omega_e$ we have

$$\omega_e = \sum_{j=1}^{N_e} v_j^e c_j^e \tag{5.19}$$

Now each entry on the right corresponds to a scaled column of the element matrix A_e. For example, for the bilinear element with four nodes, the vectors c_j are of dimension four with components $B(\psi_i, \psi_j)$, where $B(\cdot, \cdot)$ is the bilinear functional for the variational problem and ψ_k are the element basis functions. That is, for given j, $B(\psi_i, \psi_j)$ specifies the vector c_j^e. Thus it is not necessary to form A_e and then carry out the matrix vector product, but instead one can combine these two steps to a form of matrix-free scheme. The integrations in $B(\psi_i, \psi_j)$ are made on the reference element using numerical quadrature.

Conclusions

   In the past, subdomain solution techniques such as substructure analysis were introduced primarily because of resource limitations such as storage limits for large-scale problems. However, domain decomposition is a logical "divide and conquer" strategy for exploiting parallel processing. By appropriately selecting subdomain sizes and the number of subdomains per processor, good load balance and parallel efficiency can be ensured. For general geometries and graded meshes, good algorithms for subdomain specification and load balancing are needed. This problem is further complicated when local adaptive refinement of unstructured grids is permitted. In this case the subdomains should be dynamically defined between successive grid refinement-solution steps.

   Many domain-decomposition strategies are essentially equivalent to block iteration procedures and hence the definition of the subdomains need not be so explicit. For example, the subdomains are themselves unions of elements or cells in the discretized problem and, these elements need not be contiguous to define an individual subdomain. For example, a random collection of elements could be taken to define such an abstract subdomain. Similarly, a red-black type ordering of elements would give another more structured subdomain description. Of course, the parallel efficiency of the algorithm and its convergence would be affected by the choice of abstract subdomain.

The element-by-element gradient iterative schemes are appealing for parallel and vector-parallel processing with both coarse and fine granular architectures. They permit the option of parallelizing over individual elements or of grouping elements. As we have seen in this chapter, adaptive refinement can be accommodated. Both shared and distributed memory systems can be utilized, but some care must be exercised in the implementation of the algorithms in each case. For example, communication costs will be reduced if elements are grouped to contiguous subdomains and block adjacency minimize the communication path. We can anticipate further major advances in this area and particularly for the iterative algorithms considered here as the technology and software mature.

Acknowledgments

Research related to this chapter has been supported in part by the U.S. Department of Energy and the National Science Foundation. I would also like to express my appreciation to E. Barragy, S. Bova, A. Lorber, R. McLay and M. Sharma for related discussions.

References

[1] Adams, L., "Reordering Computations for Parallel Execution", *Comm. Appl. Numer. Methods*, 2, 3, 263–372, 1986.

[2] Adams, L., and R. Voigt, "A Methodology for Exploiting Parallelism in the Finite Element Process", in *Proc. of the NATO Workshop on High Speed Computation*, J. Kowalik (ed.), NATO ASI Series, Vol. F-7, Springer-Verlag, Berlin, 1984.

[3] Bank, R. E., and A. H. Sherman, *A Refinement Algorithm and Dynamic Data Structure for Finite Element Meshes*, Report TR-166, Center for Numerical Analysis, University of Texas at Austin, 1980.

[4] Barragy, E., and G. F. Carey, "A Parallel Element-by-Element Solution Scheme," *Int. J. Num. Meth. Eng.,* 26, 2367–2382, 1988.

[5] Carey, G. F., "High Speed Processors and Implications for Algorithms and Methods", in *Finite Element Methods for Nonlinear Problems in Mechanics,* W. Wunderlich (ed.), Springer Lecture Notes, 758–777, 1981.

[6] Carey, G. F., "Parallelism in Finite Element Modelling", *Comm. Appl. Numer. Methods,* 2, 3, 281–288, 1986.

[7] Carey, G. F., E. Barragy, R. McLay and M. Sharma, "Element-by-Element Vector and Parallel Computations," *Comm. Appl. Num. Meth.,* 4, 299–307, 1988.

[8] Carey, G. F., and B. N. Jiang, "Element-by-Element Linear and Nonlinear Solution Schemes", *Comm. Appl. Numer. Methods,* 2, 2, 145–153, 1986.

[9] Carey, G. F., M. Sharma and K. C. Wang, "A Class of Data Structures for 2-D and 3-D Adaptive Mesh Refinement," *Int. J. Num. Meth. Eng.,* 1989.

[10] Duff, I. S., "Parallel Implementation of Multifrontal Schemes", *Parallel Computing,* 3, 3, 193–304, 1986.

[11] Fletcher, R., "Conjugate Gradient Methods for Indefinite Systems", in *Lecture Notes in Mathematics,* Vol. 506, Springer-Verlag, 73–89, 1976.

[12] Glowinski, R., Q.V. Dinh and J. Periaux, "Domain Decomposition for Elliptic Problems", in *Finite Elements in Fluids,* Vol. 5, R.H. Gallagher *et al.* , (eds.), Wiley, New York, 1985.

[13] Glowinski, R., G. H. Golub, G. A. Meurant and J. Périaux (eds.), *Domain Decomposition Method for Partial Differential Equations,* SIAM, 1988.

[14] Hageman, L., and D. Young, *Applied Iterative Methods,* Academic Press, New York, 1981.

[15] Hayes, L. J., "Advances and Trends in Element-by-Element Techniques", in *State-of-the-Art Surveys on Computational Mechanics,* A. Noor, (ed.), ASME Winter Annual Meeting, Anaheim, CA, 1986.

[16] Hughes, T. J. R., I. Levit and J. Winget, "Element-by-Element Solution Algorithm for Problems of Structural and Solid Mechanics", *Comp. Meth. Appl. Mech. Eng.,* 36, 241–254, 1983.

[17] Ortega, J. M., and R. G. Voigt, "Solution of Partial Differential Equations on Vector and Parallel Computers", *ICASE Report 85-1*, 1985.

[18] Schwarz, H. A., "Über einige Abbildungsaufgaben," *Ges. Math. Abh.*, 11, 65–83, 1869.

[19] Sharma, M., and G. F. Carey, "Semiconductor Device Simulation Using Adaptive Refinement and Flux Upwinding," *IEEE Trans in Computer Aided Design of Integrated Circuits and Systems*, (in press), 1989.

Chapter 6

Pseudo-Boundary Conditions to Accelerate Parallel Schwarz Methods

*Garry Rodrigue** and Shantilal Shah*[†]

6.1 Domain Decomposition Methods

Domain Decomposition methods have become a popular numerical technique for obtaining parallelism when solving a partial differential equation (PDE), (see [5,1] for several papers on the subject). Briefly, the basic idea behind these methods is as follows: First one subdivides the domain of definition of the PDE into a union of subdomains and then the numerical solution of related PDE problems are obtained on each of the subdomains. These "sub-solutions" are then pieced together in some manner to obtain an approximation to the global solution.

For elliptic PDE's, the above idea becomes relatively simple since they generally revolve around an idea first developed by Schwarz [14], in 1870. He introduced an alternating procedure for solving Poisson's equation

$$-\Delta u = f(x,y), \quad (x,y) \in \Omega,$$
$$u(x,y) = g(x,y) \quad, \quad (x,y) \in \Gamma = \text{Boundary } (\Omega)$$

(6.1)

as follows:

---
*University of California, Davis, CA.
[†]Norfolk State University, Norfolk, VA.

(i) Let $\Omega = \Omega_1 \bigcup \Omega_2, \Gamma_i = $ Boundary (Ω_i) , $i = 1, 2$;

(ii) Let $u_2^{(0)}$ be an initial guess to u on Ω_2;

(iii) Define the solutions $u_1^{(k)}, u_2^{(k)}, k = 1, 2, \ldots$ on Ω_1 and Ω_2, respectively, to satisfy:

$$-\Delta u_1^{(k)} = f, \text{ on } \Omega_1,$$

$$u_1^{(k)} |_\Gamma = g,$$

$$u_1^{(k)} |_{\Gamma_1 - \Gamma} = u_2^{(k-1)} |_{\Gamma_1 - \Gamma};$$

and   (6.2)

$$-\Delta u_2^{(k)} = f, \text{ on } \Omega_2,$$

$$u_2^{(k)} |_\Gamma = g,$$

$$u_2^{(k)} |_{\Gamma_2 - \Gamma} = u_1^{(k)} |_{\Gamma_2 - \Gamma} .$$

This procedure has been termed the Schwarz Alternating Procedure. The convergence of the procedure was established by Sobolev [15], in 1936 with further improvements by Mikhlin [7,8]. The above alternating procedure (6.2) can be generalized in the obvious way to obtain a "parallel" Schwarz method by simply solving (6.2) simultaneously using boundary data from the previous iterates. Evans *et al.* [2,3,4], Tang [16] and Rodrigue *et al.* [12,13] analyzed the convergence of this and other parallel variants of the Schwarz method.

The Schwarz Alternating Procedure and many of its parallel variants are equivalent to block-Gauss-Siedel and block-Jacobi matrix iterative methods (see [6,9]). Consequently, these methods can suffer from slow convergence (albeit they are parallel). Thus, an active area of research in parallel numerical algorithms is in the development of acceleration techniques for these Schwarz methods. Because of the relationship of the Schwarz methods and matrix iterations, the popular conjugate gradient or multi-grid techniques may be used to obtain acceleration. (See [10] and [17], for example.) These approaches use the basic idea put forth by Schwarz, recast it into another form employing matrix iteration, and then apply well-known acceleration techniques, such as, coarse grid approximations. In this chapter, we re-examine the basic formulation put forth by Schwarz with the goal of redefining it to achieve a faster iteration. We will see that this redefinition does not destroy the inherent parallelism of the Schwarz methods nor does it preclude the possibility of applying any of the conjugate-gradient or coarse grid acceleration techniques that have already been successful on the classical Schwarz methods.

Looking at the basic iteration (6.2), we see that along the "pseudo-boundaries" $\Gamma_i - \Gamma, i = 1, 2$, Dirichlet boundary data is used. This leads to the question as to whether it is possible to use different boundary conditions with the hope of achieving faster convergence. That is, if $\alpha_i, \beta_i, i = 1, 2$, are real constants, we consider the following variation of (6.2):

$$
\begin{aligned}
-\Delta u_1^{(k)} &= f, \text{ on } \Omega_1, \\
u_1^{(k)}|_\Gamma &= g, \\
\alpha_1 u_1^{(k)} + \beta_1 \frac{\partial u_1^{(k)}}{\partial n}\Big|_{\Gamma_1 - \Gamma} &= \alpha_1 u_2^{(k-1)} + \beta_1 \frac{\partial u_2^{(k-1)}}{\partial n}\Big|_{\Gamma_1 - \Gamma}\ ; \\
-\Delta u_2^{(k)} &= f, \text{ on } \Omega_2, \\
u_2^{(k)}|_\Gamma &= g, \\
\alpha_2 u_2^{(k)} + \beta_2 \frac{\partial u_2^{(k)}}{\partial n}\Big|_{\Gamma_2 - \Gamma} &= \alpha_2 u_1^{(k-1)} + \beta_2 \frac{\partial u_1^{(k-1)}}{\partial n}\Big|_{\Gamma_2 - \Gamma}\ .
\end{aligned}
\tag{6.3}
$$

where n refers to the normal outward direction. Since (6.3) constitutes a new subdomain iterative process, questions of convergence and rates of convergence for both the two-domain and the multi-domain situations must be addressed. In addition, comparisons with the standard iteration (6.2) must be made. In the following section we address these issues for the solution of Poisson's equation.

6.2 Convergence Analysis

Let us consider the problem

$$
\begin{aligned}
-\Delta u &= f, && \text{on } \Omega = \{(x, y) : 0 < x, y < 1\}, \\
u|_\Gamma &= g,
\end{aligned}
\tag{6.4}
$$

where f and g are known continuous functions. The domain Ω is divided into two overlapping subdomains (Figure 6.1)

$$
\begin{aligned}
\Omega_1 &= \{(x, y) : 0 < x < x_k\ , \quad 0 < y < 1\}, \\
\Omega_2 &= \{(x, y) : x_\ell < x < 1\ , \quad 0 < y < 1\},
\end{aligned}
\tag{6.5}
$$

where $0 < x_\ell < x_k < 1$.

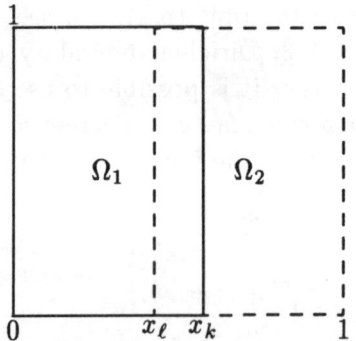

Figure 6.1: Overlapping Pair of Subdomains.

For the iteration (6.3), define

$$
\begin{aligned}
e_1^{(k)} &= u - u_1^{(k)}, &&\text{on } \Omega_1, \\
e_2^{(k)} &= u - u_2^{(k)}, &&\text{on } \Omega_2.
\end{aligned}
\tag{6.6}
$$

Then,

$$
\begin{aligned}
-\Delta e_1^{(k)} &= 0, \quad \text{on } \Omega_1, \\
e_1^{(k)}\big|_\Gamma &= 0, \\
\alpha_1 e_1^{(k)} + \beta_1 \left.\frac{\partial e_1^{(k)}}{\partial x}\right|_{\Gamma_1-\Gamma} &= \alpha_1 e_2^{(k-1)} + \beta_1 \left.\frac{\partial e_2^{(k-1)}}{\partial x}\right|_{\Gamma_1-\Gamma} \quad ; \\
-\Delta e_2^{(k)} &= 0, \quad \text{on } \Omega_2, \\
e_2^{(k)}\big|_\Gamma &= 0, \\
\alpha_2 e_2^{(k)} + \beta_2 \left.\frac{\partial e_2^{(k)}}{\partial x}\right|_{\Gamma_2-\Gamma} &= \alpha_2 e_1^{(k-1)} + \beta_2 \left.\frac{\partial e_1^{(k-1)}}{\partial x}\right|_{\Gamma_2-\Gamma} \quad .
\end{aligned}
\tag{6.7}
$$

Introducing Fourier grid expansions for the initial error,

$$
\begin{aligned}
e_1^{(0)}\big|_{\Gamma_1-\Gamma} &= \sum_{m=1}^{\infty} a_m^{(0)} \sin m\pi y \\
e_2^{(0)}\big|_{\Gamma_2-\Gamma} &= \sum_{m=1}^{\infty} b_m^{(0)} \sin m\pi y.
\end{aligned}
\tag{6.8}
$$

into (6.7), the solutions are [11],

$$
e_1^{(k)} = \sum_{m=1}^{\infty} a_m^{(k)} \frac{sh[\sigma_m x]}{sh[\sigma_m x_k]} \sin \sigma_m y, \quad \text{on } \Omega_1,
\tag{6.9}
$$

$$e_2^{(k)} = \sum_{m=1}^{\infty} b_m^{(k)} \frac{sh[\sigma_m(1-x)]}{sh[\sigma_m(1-x_\ell)]} \sin \sigma_m y, \quad \text{on } \Omega_2, \tag{6.10}$$

where $\sigma_m = m\pi$, $sh[z] = \frac{e^z - e^{-z}}{2}$, $ch[z] = \frac{e^z + e^{-z}}{2}$, and

$$a_m^{(k)} = Q(m, x_\ell, x_k) b_m^{(k-1)}, \quad b_m^{(k)} = S(m, x_\ell, x_k) a_m^{(k-1)},$$

$$Q(m, x_\ell, x_k) = \left[\frac{sh[\sigma_m x_k]}{sh[\sigma_m(x_\ell - 1)]}\right] \times$$

$$\left[\frac{\alpha_1 sh[\sigma_m(x_k - 1)] + \beta_1 \sigma_m ch[\sigma_m(x_k - 1)]}{\alpha_1 sh[\sigma_m x_k] + \beta_1 \sigma_m ch[\sigma_m x_k]}\right],$$

$$S(m, x_\ell, x_k) = \left[\frac{sh[\sigma_m(x_\ell - 1)]}{sh[\sigma_m x_k]}\right] \times$$

$$\left[\frac{\alpha_2 sh[\sigma_m x_\ell] + \beta_2 \sigma_m ch[\sigma_m x_\ell]}{\alpha_2 sh[\sigma_m(x_\ell - 1)] + \beta_2 \sigma_m ch[\sigma_m(x_\ell - 1)]}\right].$$

It follows that, for $k = 0, 1, 2, \ldots$

$$a_m^{(k+1)} = \rho(m, x_\ell, x_k) a_m^{(k-1)}, \quad b_m^{(k+1)} = \rho(m, x_\ell, x_k) b_m^{(k-1)}, \tag{6.11}$$

where $\rho(m, x_\ell, x_k) = Q(m, x_\ell, x_k) \, S(m, x_\ell, x_k)$.

Consequently, if $|\rho(m, x_\ell, x_k)| \leq \rho < 1$ for all m, then the L^2-norm of the error satisfies

$$\left\| e_1^{(k+1)} \right\|_0^2 = \sum_{m=1}^{\infty} \left| a_m^{(k+1)} \right|^2$$

$$= \sum_{m=1}^{\infty} |\rho(m, x_\ell, x_k)|^2 \left| a_m^{(k-1)} \right|^2$$

$$\leq \rho^2 \sum_{m=1}^{\infty} \left| a_m^{(k-1)} \right|^2$$

$$= \rho^2 \left\| e_1^{(k-1)} \right\|_0^2.$$

That is, for $k = 2r - 1$,

$$\left\| e_1^{(k+1)} \right\|_0 \leq \rho^r \left\| e_1^{(0)} \right\|_0.$$

Similarly,

$$\left\| e_2^{(k+1)} \right\|_0 \leq \rho^r \left\| e_2^{(0)} \right\|_0.$$

Note that if $\alpha_i = 1, \beta_i = 0, i = 1, 2$, then we have

$$\rho(m, x_\ell, x_k) = \frac{sh[\sigma_m(x_k - 1)]}{sh[\sigma_m(x_\ell - 1)]} \frac{sh[\sigma_m x_\ell]}{sh[\sigma_m x_k]} . \tag{6.12}$$

In this situation, the pseudo-boundary conditions are both Dirichlet type and the quantity in (6.12) is denoted by $\rho_{DD}(m, x_\ell, x_k)$. A simple calculation yields

$$0 < \rho_{DD}(m, x_\ell, x_k) = \frac{ch[\sigma_m(1 - d)] - ch[\sigma_m(1 - a)]}{ch[\sigma_m(1 + d)] - ch[\sigma_m(1 - a)]} \tag{6.13}$$

where $d = x_k - x_\ell$, $a = x_k + x_\ell$. Since the hyperbolic cosine is an increasing function, for $d > 0$

$$\rho_{DD}(m, x_\ell, x_k) < 1 . \tag{6.14}$$

Similarly, if $\alpha_i = 0$, $\beta_i = 1$, $i = 1, 2$, this corresponds to Neumann pseudo-boundary conditions and

$$
\begin{aligned}
\rho_{NN}(m, x_\ell, x_k) &= \frac{ch[\sigma_m(x_k - 1)]}{ch[\sigma_m(x_\ell - 1)]} \frac{ch[\sigma_m x_\ell]}{ch[\sigma_m x_k]} \\
&= \frac{ch[\sigma_m(1 - d)] + ch[\sigma_m(1 - a)]}{ch[\sigma_m(1 + d)] + ch[\sigma_m(1 - a)]} \\
&< 1 \quad (d > 0) .
\end{aligned}
\tag{6.15}
$$

Continuing, if we let $\alpha_2 = \beta_1 = 1$ and $\alpha_1 = \beta_2 = 0$, then we have a Dirichlet condition on the pseudo-boundary $\{(x_\ell, y), 0 \le y \le 1\}$ and a Neumann condition of the pseudo-boundary $\{(x_k, y), 0 \le y \le 1\}$. Moreover,

$$
\begin{aligned}
|\rho_{DN}(m, x_\ell, x_k)| &= \left| \frac{ch[\sigma_m(x_k - 1)]}{ch[\sigma_m x_k]} \frac{sh[\sigma_m x_\ell]}{sh[\sigma_m(x_\ell - 1)]} \right| \\
&= \frac{ch[\sigma_m(1 - x_k)]}{ch[\sigma_m x_k]} \frac{sh[\sigma_m x_\ell]}{sh[\sigma_m(1 - x_\ell)]} \\
&= \frac{sh[\sigma_m(1 - d)] + sh[\sigma_m(1 - a)]}{sh[\sigma_m(1 + d)] + sh[\sigma_m(1 - a)]} \\
&< 1 \quad (d > 0).
\end{aligned}
\tag{6.16}
$$

where the last inequality follows from the fact that the hyperbolic sine is an increasing function.

Now, in order to provide some indication as to which of the above three situations yields a faster algorithm, we compare the quantities (6.13), (6.15), and (6.16). To do this, first let

$$\tau(x,y) = \frac{ch[x]}{ch[y]} - \frac{sh[x]}{sh[y]} = \frac{sh[y-x]}{ch[y]sh[y]}$$

Then,

$$\tau(x,y) > 0, \text{if } y - x > 0, \text{ and} \tag{6.17}$$

$$\tau(x,y) < 0, \text{if } y - x < 0.$$

We see from (6.13) and (6.15) that

$$\rho_{DD}(m, x_\ell, x_k) < \rho_{NN}(m, x_\ell, x_k). \tag{6.18}$$

Furthermore, by (6.12) and (6.16),

$$|\rho_{DN}| - \rho_{DD} = \frac{sh[\sigma_m x_\ell]}{sh[\sigma_m(1 - x_\ell)]} \tau(\sigma_m(1 - x_k), \sigma_m x_k).$$

Consequently,

$$|\rho_{DN}| - \rho_{DD} < 0, \text{ if } x_k < \frac{1}{2}, \text{ and} \tag{6.19}$$

$$|\rho_{DN}| - \rho_{DD} > 0, \text{ if } x_k > \frac{1}{2}.$$

Thus, we see that algorithms (DD), (NN), and (DN) converge and, in general, algorithm (DD) will be faster than algorithm (NN) and algorithm (DN) will be better than algorithm (DD) for $x_k < \frac{1}{2}$. We test this result in the next section.

6.3 Computational Experiment

As a test problem, consider

$$-\Delta u = 0, \ (x,y) \in \Omega = [0,1] \times [0,1], \tag{6.20}$$

with

$$u(x,0) = 0, \quad u(x,1) = x, \quad \text{for } 0 \le x \le 1, \text{ and}$$
$$u(0,y) = 0, \quad u(1,y) = y, \quad \text{for } 0 \le y \le 1. \tag{6.21}$$

In this case, the exact solution is $u = xy$.

A uniform grid is placed on Ω with $\Delta x = \Delta y = \frac{1}{16}$. Algorithm (6.3) will be used to solve (6.20) where the "Δ" operators are replaced by central difference operators. The domain decomposition of Ω is given by (6.5) where $x_k = i\Delta x$, $x_\ell = j\Delta x$ for some integers $j < i$. The individual sub-problems in (6.3) are solved to an error tolerance of 10^{-11} using the Preconditioned Conjugate Gradient method with a point-Jacobi preconditioner. In Table 6.1, we compare the number of iterations for algorithms (DD) and (DN) to achieve $\| u_{\text{computed}} - u_{\text{exact}} \|_\infty < 10^{-11}$. In all cases $d = x_k - x_\ell$.

Table 6.1: Convergence of DD and DN Schemes.

	Iterations					
	$d = .125$		$d = .1875$		$d = .25$	
j	DD	DN	DD	DN	DD	DN
2	43	23	32	20	22	18
4	52	35	36	29	27	24
6	51	44	37	34	29	28
8	53	52	37	39	28	32
10	50	63	33	45	25	37
12	41	70	25	63		

Note that on a parallel computer, best efficiency is obtained when the size of each subdomain is the same and the overlap of the subdomain is minimal so that efficiency is achieved when $j = 6$ and $d = .125$ in the present case. Table 6.1 reveals that algorithm (DN) is then preferred.

6.4 Three-Subdomains

The 3-subdomain case in Figure 6.2 is now investigated. Equation (6.20) is to be solved via an extension of (6.3) on the three domains $\Omega_1 = [0, x_{k_1}] \times [0, 1]$, $\Omega_2 = [x_{\ell_2}, x_{k_2}] \times [0, 1]$, $\Omega_3 = [x_{\ell_3}, 1] \times [0, 1]$ where $0 < x_{\ell_2} < x_{k_1} < x_{\ell_3} < x_{k_2} < 1$ (Figure 6.2). On the partitioning, we consider the following iteration:

a)
$$-\Delta u_1^{(k)} = f, \qquad \text{on } \Omega_1,$$

$$u_1^{(k)}|_{\Gamma} = g,$$

$$\alpha_1 u_1^{(k)} + \beta_1 \frac{\partial u_1^{(k)}}{\partial x}\Big|_{\Gamma_1 - \Gamma} = \alpha_1 u_2^{(k-1)} + \beta_1 \frac{\partial u_2^{(k-1)}}{\partial x}\Big|_{\Gamma_1 - \Gamma};$$

b)
$$-\Delta u_2^{(k)} = f, \qquad \text{on } \Omega_2,$$

$$u_2^{(k)}|_{\Gamma} = g,$$

$$\alpha_2 u_2^{(k)} + \beta_2 \frac{\partial u_2^{(k)}}{\partial x}\Big|_{(\Gamma_2 - \Gamma) \bigcap \Omega_1} = \alpha_2 u_1^{(k-1)} + \beta_2 \frac{\partial u_1^{(k-1)}}{\partial x}\Big|_{(\Gamma_2 - \Gamma) \bigcap \Omega_1},$$

$$\alpha_3 u_2^{(k)} + \beta_3 \frac{\partial u_2^{(k)}}{\partial x}\Big|_{(\Gamma_2 - \Gamma) \bigcap \Omega_3} = \alpha_3 u_3^{(k-1)} + \beta_3 \frac{\partial u_3^{(k-1)}}{\partial x}\Big|_{(\Gamma_2 - \Gamma) \bigcap \Omega_3};$$

c)
$$-\Delta u_3^{(k)} = f, \qquad \text{on } \Omega_3,$$

$$u_3^{(k)}|_{\Gamma} = g,$$

$$\alpha_4 u_3^{(k)} + \beta_4 \frac{\partial u_3^{(k)}}{\partial x}\Big|_{\Gamma_3 - \Gamma} = \alpha_4 u_2^{(k-1)} + \beta_4 \frac{\partial u_2^{(k-1)}}{\partial x}\Big|_{\Gamma_3 - \Gamma}.$$

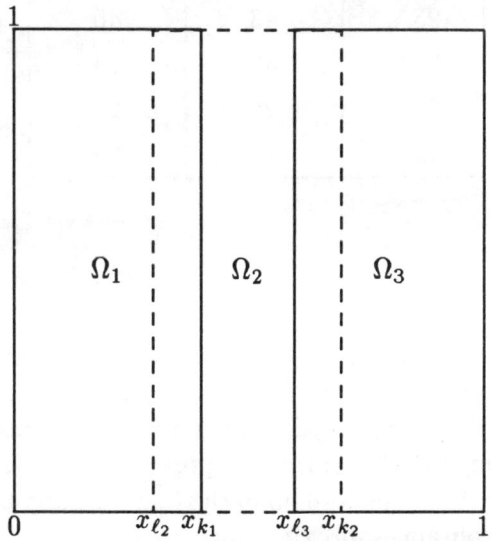

Figure 6.2: Three Subdomains.

Specifically, the classical situation with Dirichlet conditions on the pseudo-boundaries, defined by

$$(DD) \qquad\qquad \alpha_i = 1, \quad \beta_i = 0, \qquad i = 1,2,3,4$$

is to be compared with the mixed Neumann-Dirichlet situation

$$(DN) \qquad \alpha_2 = \beta_1 = \beta_3 = \alpha_4 = 0, \quad \beta_2 = \alpha_1 = \alpha_3 = \beta_4 = 1 \; .$$

The computational parameters are the same as in Section 6.3.

In the numerical experiment, we let $x_{k(1)} = 7\Delta x$ and $x_{\ell(3)} = 9\Delta x$ be fixed and

$$\mid x_{k(2)} - 8\Delta x \mid \; = \; \mid x_{\ell(2)} - 8\Delta x \mid = d$$

be allowed to vary. The number of iterations necessary for convergence of algorithms (DD) and (DN) above are recorded in Table 6.2. Again we see that algorithm (DN) is faster for the optimal parallel situation of equally-sized subdomains and minimal overlap (i.e $d = .1875$).

Table 6.2: Iterative Performance.

d	DD	DN
.1875	69	49
.25	43	40
.3125	33	34
.375	25	29
.4375	19	26

6.5 Conclusions

Partitioning the domain of an elliptic partial differential equation is an effective starting point for obtaining parallel numerical tasks. In defining the subproblems on these subdomains, boundary conditions are needed along interior "pseudo-boundaries". Current approaches have been to guess initial Dirichlet data along the "pseudo-boundaries" and iteratively update between the subdomains to obtain approximations to the global solution. In this chapter, we have shown that mixed boundary conditions may provide a more efficient iterative scheme. The resulting schemes parallelize as in the previous case.

References

[1] Chan, T., (ed.), *Proceedings Second International Symposium on Domain Decomposition Methods for for Partial Differential Equations*, SIAM Publications, Philadelphia, 1988.

[2] Evans, D., and L. Kang., "The Convergence Rate of the Schwarz Alternating Procedure (III)—for Neumann Problems," *Int. J. Comput. Math.*, 21, 85–108, 1987.

[3] Evans, D., L. Kang and J. Shao, "The Convergence Rate of the Schwarz Alternating Procedure (I)— for One-dimensional Problems," *Int. J. Comput. Math.*, 20, 157–170, 1986.

[4] Evans, D., L. Kang, J. Shao and Y. Chen, "The Convergence Rate of the Schwarz Alternating Procedure (II)—for Two-Dimensional Problems," *Int. J. Comput. Math.*, 20, 325–339, 1986.

[5] Glowinski, R., G. Golub, G. Meurant and J. Periaux, (eds.), *Proceedings First International Symposium on Domain Decomposition Methods for Partial Differential Equations*, SIAM Publications, Philadelphia, 1987.

[6] Kang, L., *Domain Decomposition and Parallel Algorithms*, Wuhan University Press, Wuhan University, Peoples Republic of China, 1987.

[7] Mikhlin, S. G., "On the Schwarz Algorithm," *DAH CCCP*, 77, 4, 569–571, 1951.

[8] Mikhlin, S. G., *The Problem of the Minimum of a Quadratic Functional*, Holden-Day Publishers, 1965.

[9] Rodrigue, G., "Domain Decomposition and Inner/Outer Iteration for Elliptic Partial Differential Equations," *J. Parallel Computing*, 2, 205–218, 1985.

[10] Rodrigue, G. and T. Ferretta, "Coarse Grid Acceleration of Some Domain Decomposition Methods on Multiprocessors," *Proceedings IFIPS Conference*, Stanford, Cal., 1988, (also available as Lawrence Livermore Laboratory Report, UCRL-99762, Oct., 1988).

[11] Rodrigue, G., L. Kang, G. Lin and Z. Wu, "The Generalized Schwarz Alternating Principle," Lawrence Livermore Laboratory Report, UCRL-96770, Sept., 1987.

[12] Rodrigue, G., and Y. Liu, *Convergence and Comparison Analysis of Some Numerical Schwarz Methods*, Lawrence Livermore Laboratory Report, UCRL-98885, May, 1988.

[13] Rodrigue, G., and J. Simon, "Jacobi Splittings and the Method of Overlapping Domains for Solving Elliptic P.D.E.'s.," *Proceedings of the 5th IMACS International Symposium*, Bethlehem, Pa., June, 1984.

[14] Schwarz, H. A., "Uber Enen Grenz Bergang Durich Alternirender Verfahren," *Ges. Math. Abhandlungen*, Bd. 1, Berlin, 133–143, 1870.

[15] Sobolev, S., "L'Algorithme de Schwarz dans la Theorie de l'Elasticite," *Compte Rendus (Doklady) de l'Accidemie des Sciences de L'URSS*, Vol. IV (VIII), 6(110), 243–246, 1936.

[16] Tang, W. P., *Schwarz Splitting and Template Operators*, Ph.D. Dissertation, Dept. of Computer Science, Stanford Univ., 1987.

[17] Widlund, O., *Some Domain Decomposition and Iterative Refinement Algorithms for Elliptic Finite Element Problems*, CS Technical Report No. 386, Courant Institute of Mathematical Sciences, N.Y. Univ., 1988.

Chapter 7

The Search for "High-Level" Parallelism for Iterative Sparse Linear System Solvers

*David M. Young**

7.1 Introduction

In this chapter we are concerned with the numerical solution, based on iterative methods, of large sparse systems of linear algebraic equations. Of particular interest are systems which arise in the numerical solution of elliptic and parabolic partial differential equations by finite difference or finite element methods. We consider linear systems of the form

$$Au = b \tag{7.1}$$

where A is a given $N \times N$ matrix which is large and sparse and where b is a given $N \times 1$ column vector. We will assume that A is symmetric and positive definite (SPD). We consider iterative algorithms for solving (7.1) which consist of a "basic iterative method," such as the Richardson, Jacobi, SSOR or incomplete Cholesky method, combined with an acceleration procedure such as Chebyshev acceleration or conjugate gradient acceleration. (For a discussion of these procedures, see, e.g., Hageman and Young [5].)

It is often possible to achieve parallelism for iterative algorithms by subdividing the matrix problem into blocks and assigning each processor the task

*Center for Numerical Analysis, The University of Texas, Austin, TX.

of handling one or more blocks. For problems arising from partial differential equations this corresponds to subdividing the region into subregions. Such procedures lead to block iteration and domain decomposition methods, for example. One can often greatly increase the convergence of iterative algorithms by the use of multigrid techniques wherein one focusses on the use of several grids.

The object of this chapter is, however, to examine some "high-level" methods for achieving parallelism. Such techniques involve only matrix/vector operations and do not involve working with blocks of the matrix, subdividing the region, or using different meshes. It is expected that if effective high-level methods could be developed, they could be combined with block and domain decomposition methods, and related methods, to obtain even greater speedups. It is also expected that by working at a higher level it will eventually be possible to develop general purpose software for parallel machines similar to the `ITPACK` software packages which have already been developed for sequential and vector machines; see Kincaid and Young [8].

Our discussion here is primarily devoted to describing various techniques which we and others have considered for obtaining high-level parallelism. We plan to continue research on these techniques and eventually to develop algorithms and programs for multiprocessors based on them.

In Section 7.2 we describe some "parallel iteration" techniques. Here several iteration procedures are applied in parallel and the results are combined periodically to yield (hopefully) faster convergence than that produced by any one of the individual procedures used. Similarly in Section 7.3 we consider "residual decomposition" techniques wherein an initial residual, corresponding to a given starting vector, is decomposed into the sum of several subresiduals. An iterative procedure is then applied to all of the subresiduals in parallel and the results combined to yield (hopefully) faster convergence. If the decomposition of the residual is carried out according to the eigenvalue spectrum of the matrix A then we refer to the procedure as a "spectral decomposition method."

For the methods used in both Section 7.2 and Section 7.3 several corrections to the initial approximation vector $u^{(0)}$ are obtained. A linear combination of these corrections is used in order to minimize a certain norm of the error. Such a procedure is described in Section 7.4. This procedure is related to the *conjugate direction method* which is also described in Section 7.4 and which is contrasted to the conjugate gradient method.

The discusson of Section 7.4 is then applied, in Section 7.5 to the problem of solving a family of linear systems

$$(A + \rho I)u = b \tag{7.2}$$

where the scalar ρ and the vector b may vary. It is shown how, by the use of Arnoldi vectors and the conjugate direction method, only one set of matrix/vector multiplications involving the matrix A is required to solve all of the derived systems.

In Section 7.6 we consider the time-dependent problem defined by

$$\frac{du(t)}{dt} = -Au(t) + b \tag{7.3}$$

where A is a fixed $N \times N$ SPD matrix and where b is a fixed $N \times 1$ column vector. We consider the use of the backward difference method and the Crank-Nicolsen method. Each scheme involves the repeated solution of systems of the form (7.2). Moreover it is shown that, by the use of partial fraction representations of rational functions, several time steps can be carried out in parallel provided that the time steps are of different sizes.

It is well-known that there is a close relation between the solution of time-dependent problems of the form (7.3) and "steady state" problems of the form (7.1). This suggests the use of "rational" iteration techniques, described in Section 7.7. For these techniques we have $\varepsilon^{(n)} = R_n(A)\varepsilon^{(0)}$ where $\varepsilon^{(0)}$ is the initial error vector. The function $R_n(A)$ is a rational function. For polynomial iteration, also called "polynomial acceleration," $R_n(A)$ is a polynomial in A. Rational iteration often converges very rapidly and has many other desirable properties. However, to carry out each iteration requires the solution of a linear system of the form (7.2), which may be very costly if ρ is small. It is hoped that techniques can be developed, possibly based on the use of the procedures developed in Section 7.4, to overcome this difficulty.

7.2 Parallel Iteration

Let us consider the following procedure for solving (7.1). We choose an initial approximation $u^{(0)}$ to the true solution $\bar{u} = A^{-1}b$ of (7.1). If $u^{(0)} \neq 0$ we may replace $u^{(0)}$ by $cu^{(0)}$ where c is a scalar chosen to minimize the error norm

$$\|cu^{(0)} - \bar{u}\|_{A^{1/2}} \tag{7.4}$$

where, in general, the $A^{1/2}$-norm of a vector v is given by

$$\|v\|_{A^{1/2}} = (v, Av)^{1/2} . \tag{7.5}$$

The choice of c to minimize (7.4) is

$$c = \frac{(b, u^{(0)})}{(u^{(0)}, Au^{(0)})} . \tag{7.6}$$

In the following discussion we will assume, for convenience of presentation, that $c = 1$.

The idea of parallel iteration is to carry out several iteration procedures starting with $u^{(0)}$, thus obtaining $u^{(1)}, u^{(2)}, \ldots, u^{(s)}$. One then chooses constants $c_1, c_2, \ldots, c_s$ so that $\|u - \bar{u}\|_{A^{1/2}}$ is minimized, where

$$u = \sum_{i=1}^{s} c_i u^{(i)} . \qquad (7.7)$$

A procedure for finding the c_i is given in Section 7.4. Having determined u one can repeat the process, replacing $u^{(0)}$ by u. The hope is that, by using s iterative methods in parallel, the rate of convergence of the overall procedure would be increased, ideally by a factor close to s.

Adams [1] considers "additive M-step preconditioners." Two iterative procedures are used—one based on the foward SOR method and the other based on the backward SOR method. The resulting iterative process, which could be carried out in parallel, is combined with polynomial preconditioning, [3] and [7]. The results obtained compare favorably with those obtained using the SSOR method. We note that the SSOR method can be regarded as a multiplicative, rather than an additive, preconditioning procedure since it involves a forward SOR iteration followed by a backward SOR iteration.

Very little appears to be known in general about the speedup attainable by parallel iteration. However, O'Leary and White [11] have proved some convergence results for iterative methods based on multi-splittings. Other results are given by Frommer and Mayer [4]. Research is needed to determine whether, for a given problem or class of problems, significant speedups are possible and, if so, how the iterative methods should be chosen.

7.3   Residual Decomposition

The idea of residual decomposition is somewhat similar to that of parallel iteration. Suppose we are given an initial approximation $u^{(0)}$, to the true solution $\bar{u} = A^{-1}b$ of (7.1). We decompose the residual $r^{(0)} = b - Au^{(0)}$ into s "subresiduals" $r^{(0,1)}, r^{(0,2)}, \ldots, r^{(0,s)}$ such that

$$r^{(0)} = r^{(0,1)} + r^{(0,2)} + \cdots + r^{(0,s)} . \qquad (7.8)$$

We then solve the systems

$$A\Delta^{(i)} = r^{(0,i)} , \qquad i = 1, 2, \ldots, s \qquad (7.9)$$

to obtain the corrections $\Delta^{(1)}, \Delta^{(2)}, \ldots, \Delta^{(s)}$. If (7.9) is solved exactly, then

$$\bar{u} = A^{-1}b = u^{(0)} + \Delta^{(1)} + \Delta^{(2)} + \cdots + \Delta^{(s)} . \qquad (7.10)$$

However, if the $\boldsymbol{\Delta}^{(i)}$ are solved only approximately, we choose as our new approximate solution

$$\hat{u} = u^{(0)} + c_1\tilde{\boldsymbol{\Delta}}^{(1)} + c_2\tilde{\boldsymbol{\Delta}}^{(2)} + \cdots + c_s\tilde{\boldsymbol{\Delta}}^{(s)} \tag{7.11}$$

where, for each i, $\tilde{\boldsymbol{\Delta}}^{(i)}$ is an approximate solution of (7.9). We choose $c_1, c_2, \ldots, c_s$ to minimize $\|\hat{u} - \bar{u}\|_{A^{1/2}}$. Procedures for choosing the $\{c_i\}$ are given in the next section.

Spectral Decomposition

Let us now consider the possibility of decomposing $r^{(0)}$ on the basis of a decomposition of the spectrum of the coefficient matrix $\boldsymbol{A}$ of (7.1). As an example, let us consider a decomposition into three parts. As in Figure 7.1 we subdivide the interval $[m(\boldsymbol{A}), M(\boldsymbol{A})]$, where $m(\boldsymbol{A})$ and $M(\boldsymbol{A})$ are, respectively, the smallest and largest eigenvalues of $\boldsymbol{A}$, into three subintervals, namely, $I_1 = [\alpha_0, \alpha_1]$, $I_2 = [\alpha_1, \alpha_2]$, and $I_3 = [\alpha_2, \alpha_3]$ where $\alpha_0 = m(\boldsymbol{A})$ and $\alpha_3 = M(\boldsymbol{A})$. We write the residual $r^{(0)}$ in the form

$$r^{(0)} = \sum_{i=1}^{N} c_i v^{(i)} \tag{7.12}$$

where $v^{(i)}$ is the eigenvector of $\boldsymbol{A}$ associated with the eigenvalue ν_i. We seek to choose $r^{(0,k)}$, $k = 1, 2, 3$ so that, if

$$r^{(0,k)} = \sum_{i=1}^{N} c_i^{(k)} v^{(i)}, \tag{7.13}$$

then all values of $c_i^{(k)}$ are small except for values of i such that $\nu_i \in I_k$. If $\nu_i \in I_k$ we desire that $c_i^{(k)} = c_i$. The advantage of this is as follows: If we use the conjugate gradient method to solve

$$A\boldsymbol{\Delta} = r^{(0)} \tag{7.14}$$

the number of iterations is of the order of $\sqrt{K(\boldsymbol{A})}$ where $K(\boldsymbol{A}) = M(\boldsymbol{A})/m(\boldsymbol{A})$ is the condition number of $\boldsymbol{A}$. On the other hand if we let

$$\alpha_1 = K(\boldsymbol{A})^{1/3} \quad , \quad \alpha_2 = K(\boldsymbol{A})^{2/3} \tag{7.15}$$

then we have

$$\frac{\alpha_3}{\alpha_2} = \frac{\alpha_2}{\alpha_1} = \frac{\alpha_1}{\alpha_0} = K(\boldsymbol{A})^{1/3} . \tag{7.16}$$

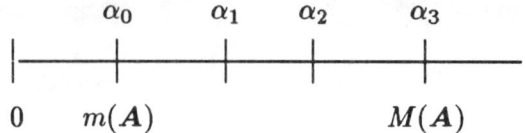

Figure 7.1: Decomposition of the Spectrum of A.

If we apply the conjugate gradient method to solve each of the systems

$$A\Delta^{(k)} = r^{(0,k)} ,\tag{7.17}$$

which can be done in parallel, then the number of iterations will be on the order of $K(A)^{1/6}$. This would be true if all of the $c_i^{(k)}$ were to vanish *exactly* for all i such that $\nu_i \notin I_k$. Thus the procedure has considerable potential.

We have developed a program for splitting $r^{(0)}$. This program is based on the construction of orthogonal polynomials using a three term relation and on the determination of the characteristic function of each of the subintervals I_1, I_2 and I_3. By operating on $r^{(0)}$ by each of these characteristic functions, which are polynomials in A, we can get fairly good subresiduals $r^{(0,1)}$, $r^{(0,2)}$ and $r^{(0,3)}$. Unfortunately however, the components $c_i^{(k)}$ are not *exactly* zero outside of the k-th interval I_k. Our numerical experiments indicate that unless the $c_i^{(k)}$ are extremely close to zero outside of I_k, then the number of iterations required is on the order of $K(A)^{1/2}$ rather than $K(A)^{1/6}$. Thus the procedure, as it stands, does not appear to be practical. We are however, continuing our research on the development of practical methods for decomposing $r^{(0)}$.

We note that rather than using residual decomposition we can choose several starting vectors, say, $u^{(0,1)}, u^{(0,2)}, \ldots, u^{(0,s)}$. We can then carry out m steps of a given iterative procedure using each of the starting vectors, obtaining, say $u^{(m,1)}, u^{(m,2)}, \ldots, u^{(m,s)}$. This can be done in parallel. We can then choose scalars $k_1, k_2, \ldots, k_s$ so that $\|u - \bar{u}\|_{A^{1/2}}$ is minimized, where

$$u = \sum_{i=1}^{s} k_i u^{(m,i)} .\tag{7.18}$$

For the case where the iterative procedure used is the conjugate gradient method this scheme is related to the block conjugate gradient procedure described by O'Leary, [9,10].

Research is needed to determine whether a significant speedup can be achieved by the use of several different starting vectors, and, if so, how the starting vectors should be chosen. One possibility for problems arising from partial differential equations might be to let the $\{u^{(0,i)}\}$ correspond to those elements of a hierarchical basis with large support. Alternatively, for the standard five-point finite difference representation of the Dirichlet problem in the unit square, for example, we might choose the comparatively "smooth" functions $\sin \pi x \sin \pi y$, $\sin 2\pi x \sin \pi y$, $\sin \pi x \sin 2\pi y$, etc.

7.4 Minimization Procedures

Let us assume that we wish to solve the system (7.1), where A is SPD and that we have an initial approximation $u^{(0)}$ to the solution vector $\bar{u} = A^{-1}b$. Let us also assume that we have $s+1$ linearly independent, "direction vectors" $v^{(0)}, v^{(1)}, \ldots, v^{(s)}$. We seek to determine a vector u^* of the form

$$u^* = u^{(0)} + \sum_{i=0}^{s} c_i v^{(i)} \tag{7.19}$$

such that $F(u^*)$ is minimized, where

$$F(u) = \|u - \bar{u}\|^2_{A^{1/2}} = (u - \bar{u}, A(u - \bar{u})) . \tag{7.20}$$

To minimize $F(u)$ we first construct a set of $s + 1$ modified direction vectors $p^{(0)}, p^{(1)}, \ldots, p^{(s)}$ which are mutually A-orthogonal, or "conjugate," in the sense that

$$(p^{(i)}, Ap^{(j)}) = 0 , \qquad i \neq j . \tag{7.21}$$

To do this we use the Gram-Schmidt procedure. We have

$$\begin{cases} p^{(0)} = v^{(0)} \\[2mm] p^{(1)} = v^{(1)} - \dfrac{(v^{(1)}, Ap^{(0)})}{(p^{(0)}, Ap^{(0)})} \, p^{(0)} \\[3mm] p^{(2)} = v^{(2)} - \dfrac{(v^{(2)}, Ap^{(0)})}{(p^{(0)}, Ap^{(0)})} \, p^{(0)} - \dfrac{(v^{(2)}, Ap^{(1)})}{(p^{(1)}, Ap^{(1)})} \, p^{(1)} \\[3mm] \cdots \end{cases} \tag{7.22}$$

We then determine u^* by

$$u^* = u^{(0)} + \sum_{k=0}^{s} k_i p^{(i)} \tag{7.23}$$

where

$$k_i = \frac{(p^{(i)}, r^{(0)})}{(p^{(i)}, Ap^{(i)})} \tag{7.24}$$

and
$$r^{(0)} = b - Au^{(0)} . \tag{7.25}$$
This follows since $A(u^{(0)} - \bar{u}) = Au^{(0)} - b = -r^{(0)}$ and since (7.21) holds.

Conjugate Direction Method

Let us again assume that we wish to solve the linear system (7.1) where A is SPD, and that we have a set of N linearly independent (direction) vectors $p^{(0)}, p^{(1)}, \ldots, p^{(N-1)}$ which are mutually A-orthogonal. (If we have N linearly independent vectors which are not mutually A-orthogonal then one can, in principle at least, obtain mutually A-orthogonal vectors by the Gram-Schmidt process described above.) The conjugate direction method can be defined by

$$
\begin{cases}
u^{(0)} & \text{is arbitrary} \\
r^{(0)} &= b - Au^{(0)} \\
u^{(n+1)} &= u^{(n)} + \lambda_n p^{(n)} , \qquad n = 0, 1, \ldots, N-1 \\
\lambda_n &= \dfrac{(p^{(n)}, r^{(0)})}{(p^{(n)}, Ap^{(n)})}
\end{cases}
\tag{7.26}
$$

We remark that the conjugate gradient method [6] is a special case of the conjugate direction method where the direction vectors are computed sequentially. Thus for the conjugate gradient method we have:

$$
\begin{aligned}
u^{(0)} \quad & \text{is arbitrary} \\
r^{(0)} &= b - Au^{(0)} \\
p^{(n)} &= r^{(n)} + \alpha_n p^{(n-1)} \\
u^{(n+1)} &= u^{(n)} + \lambda_n p^{(n)} \\
r^{(n)} &= b - Au^{(n)} \\
\alpha_n &= \frac{(r^{(n)}, r^{(n)})}{(r^{(n-1)}, r^{(n-1)})} \\
\lambda_n &= \frac{(r^{(n)}, p^{(n)})}{(p^{(n)}, Ap^{(n)})}
\end{aligned}
\tag{7.27}
$$

The direction vectors and residuals are computed in the order $r^{(0)}$, $p^{(0)}$, $r^{(1)}$, $p^{(1)}$, $\ldots$.

7.5 Solution of Related Linear Systems

Let us suppose that we wish to solve a family of linear systems of the form
(7.2). We assume that A is a fixed SPD matrix and that the nonnegative
constant ρ and the vector b may vary. If all of the ρ's and b's were known
in advance, the solutions could be obtained in parallel using the conjugate
gradient method. However, we assume that we need to solve the systems
sequentially so that ρ and b are not known in advance.

We propose the following strategy: We first choose a vector $w^{(0)}$ and
a value of ρ, say ρ_0. For a given integer, s, we construct a set of vectors
$w^{(1)}, w^{(2)}, \ldots, w^{(s)}$ called "Arnoldi vectors," which span $K_s(w^{(0)}, A + \rho_0 I) =
Sp(w^{(0)}, (A + \rho_0 I)w^{(0)}, \ldots, (A + \rho_0 I)^{s-1} w^{(0)})$ and which are mutually orthog-
onal but not, in general, mutually $(A + \rho_0 I)$-orthogonal. We then show that
the $\{w^{(i)}\}$ are independent of ρ_0. Then, for any given ρ we construct a set of
direction vectors $p^{(0)}, p^{(1)}, \ldots, p^{(s)}$ which are mutually $(A + \rho I)$-orthogonal.
The conjugate direction method is then applied to obtain an approximate
solution of (7.2).

The Arnoldi Vectors

Given $w^{(0)}$ and ρ_0 we construct the Arnoldi vectors using the formula

$$w^{(i)} = (A + \rho_0 I)w^{(i-1)} + \beta_{i,i-1} w^{(i-1)} + \beta_{i,i-2} w^{(i-2)} \qquad (7.28)$$

where

$$
\begin{cases}
\beta_{i,i-1} = -\dfrac{\left((A + \rho_0 I)w^{(i-1)}, w^{(i-1)}\right)}{(w^{(i-1)}, w^{(i-1)})} \\[4mm]
\beta_{i,i-2} = -\dfrac{\left((A + \rho_0 I)w^{(i-1)}, w^{(i-2)}\right)}{(w^{(i-2)}, w^{(i-2)})}
\end{cases}
\qquad (7.29)
$$

It can easily be shown that the $\{w^{(i)}\}$ are mutually orthogonal. It can also
be shown that the $\{w^{(i)}\}$ are independent of ρ_0 since we have

$$
\begin{cases}
\beta_{i,i-1}(\rho_0) = \beta_{i,i-1}(0) - \rho_0 \\[2mm]
\beta_{i,i-2}(\rho_0) = \beta_{i,i-2}(0)
\end{cases}
\qquad (7.30)
$$

It should be noted that only $s_0 + 1$ vectors can be obtained by the process
where s_0 is the smallest integer such that the vectors $w^{(0)}, (A + \rho_0 I)w^{(0)}, \ldots,$
$(A + \rho_0 I)^{s_0} w^{(0)}$ are linearly dependent. Evidently $s_0 \leq N - 1$.

The Direction Vectors

Let us now construct the direction vectors $p^{(0)}, p^{(1)}, \ldots, p^{(s)}$ corresponding to a given value of ρ which will differ from ρ_0. We define $p^{(0)}, p^{(1)}, \ldots$ by

$$
\left\{
\begin{aligned}
p^{(0)} &= w^{(0)} \\
p^{(1)} &= w^{(1)} + a_1 p^{(0)} \\
p^{(2)} &= w^{(2)} + a_2 p^{(1)} \\
&\cdots
\end{aligned}
\right.
\tag{7.31}
$$

where

$$
a_n = -\frac{(w^{(n)}, (A + \rho I)p^{(n-1)})}{(p^{(n-1)}, (A + \rho I)p^{(n-1)})}
\tag{7.32}
$$

We first show that $(w^{(i)}, p^{(j)}) = 0$ for $j < i$. But $p^{(j)}$ is a linear combination of $w^{(0)}, w^{(1)}, \ldots, w^{(j)}$ so that the result follows from the orthogonality of the $\{w^{(i)}\}$. We next show that $(p^{(n)}, (A + \rho I)p^{(i)}) = 0$ for $i = 0, 1, \ldots, n-1$. This is true for $i = n-1$ by (7.32). For $i \leq n-2$ we have

$$
\left(p^{(0)}, (A + \rho I)p^{(i)} \right) = \left(w^{(n)} + a_n p^{(n-1)}, (A + \rho I)p^{(i)} \right)
$$

$$
= \left(w^{(n)}, (A + \rho I)p^{(i)} \right)
\tag{7.33}
$$

since $i < n-1$. But $p^{(i)}$ is a linear combination of $w^{(i)}, w^{(i-1)}, \ldots, w^{(0)}$. Hence $(A + \rho I)w^{(i)}$ is a linear combination of $w^{(i+1)}, w^{(i)}, \ldots, w^{(0)}$. Since $i < n-1$ the result follows from the orthogonality of the $\{w^{(i)}\}$.

Suppose now that we wish to solve a specific linear system (7.20 and that we have already computed and stored the Arnoldi vectors as well as the $\{(A + \rho_0 I)w^{(i)}\}$. We use the conjugate direction method with direction vectors given by (7.31). We show that this does not require any additional matrix/vector multiplications involving A. This is possible since by (7.31) we have

$$
(A + \rho I)p^{(i)} = (A + \rho I)w^{(i)} + (A + \rho I)a_i p^{(i-1)} .
\tag{7.34}
$$

Thus we can compute $(A + \rho I)p^{(0)}, (A + \rho I)p^{(1)}$, etc. recursively using $(A + \rho I)w^{(0)}, (A + \rho I)w^{(1)}$, etc. (We note that $(A + \rho I)w^{(i)} = (A + \rho_0 I)w^{(i)} + (\rho - \rho_0)w^{(i)}$.)

For a given initial approximation $u^{(0)}$ to the true solution $\bar{u}$ we compute $r^{(0)} = b - (A + \rho I)u^{(i)}$ and $u^{(1)}, u^{(2)}, \ldots$ by

$$
u^{(n+1)} = u^{(n)} + \lambda_n p^{(n)} , \qquad n = 0, 1, 2, \ldots
\tag{7.35}
$$

where

$$\lambda_n = \frac{(p^{(n)}, r^{(0)})}{(p^{(n)}, (A + \rho I)p^{(n)})} \tag{7.36}$$

We remark that the above scheme may break down if only s_0 linearly independent Arnoldi vectors $\{w^{(i)}\}$ can be generated and if the conjugate direction method does not yield sufficient accuracy within s_0 iterations. If this happens one could generate a new set of Arnoldi vectors based on $r^{(0)} = b - (A + \rho I)u^{(0)}$. However, the savings of the matrix/vector multiplication by $A + \rho I$ would be lost.

7.6 Time Dependent Problems

Let us now consider the time dependent problem (7.3). Such a problem arises, for example, from the standard five-point difference equation representation, with respect to the space variables, of a problem involving the diffusion equation

$$\frac{\partial u}{\partial t} = \frac{\partial^2 u}{\partial x^2} + \frac{\partial^2 u}{\partial y^2} + f(x, y) \tag{7.37}$$

over a rectangle, where the values of u are given and fixed for all t on the boundary of the rectangle and the initial distribution of u is given.

We consider two alternative discretizations for solving (7.3). The first corresponds to the so-called *backward difference method* defined by

$$\frac{u(t + \Delta t) - u(t)}{\Delta t} = -Au(t + \Delta t) + b \tag{7.38}$$

or

$$(A + \rho I)u(t + \Delta t) = \rho u(t) + b \tag{7.39}$$

where

$$\rho = \frac{1}{\Delta t} . \tag{7.40}$$

The second corresponds to the *Crank-Nicolson method* defined by

$$\frac{u(t + \Delta t) - u(t)}{\Delta t} = -A \left\{ \frac{u(t + \Delta t) + u(t)}{2} \right\} + b \tag{7.41}$$

or

$$(A + \rho I)u(t + \Delta t) = 2b - (A - \rho I)u(t) \tag{7.42}$$

where

$$\rho = \frac{2}{\Delta t} \tag{7.43}$$

Let us now focus on the backward difference equation (7.38). Suppose we use the time steps $(\Delta t)_1$ and $(\Delta t)_2$ where $(\Delta t)_1 \neq (\Delta t)_2$. We have, by (7.39),

$$u\left(t + (\Delta t)_1 + (\Delta t)_2\right) = (A + \rho_2 I)^{-1}(A + \rho_1 I)^{-1}\{\rho_1 \rho_2 u(t) + \rho_2 b\}$$

$$+ (A + \rho_2 I)^{-1} b . \tag{7.44}$$

We now consider the partial fraction representation of $(x + \rho_2)^{-1}(x + \rho_1)^{-1}$, which is given by

$$\frac{1}{(x + \rho_2)(x + \rho_1)} = \frac{1}{\rho_1 - \rho_2}\left\{\frac{1}{x + \rho_2} - \frac{1}{x + \rho_1}\right\} . \tag{7.45}$$

Thus we have

$$u\left(t + (\Delta t)_1 + (\Delta t)_2\right) = \frac{1}{\rho_1 - \rho_2}\left\{(A + \rho_2 I)^{-1} - (A + \rho_1 I)^{-1}\right\}$$

$$\times \ \{\rho_1 \rho_2 u(t) + \rho_2 b\} + (A + \rho_2 I)^{-1} b . \tag{7.46}$$

Evidently to carry out the above process we have to solve two systems of the form (7.2) for two sets of values of ρ and b. These systems can be solved in parallel using the techniques of the previous section. The idea can be extended to allow for several time steps $(\Delta t)_1$, $(\Delta t)_2, \ldots, (\Delta t)_s$ provided that the $\{(\Delta t)_i\}$ are distinct.

We remark that the idea of using partial fractions for the parallel solution of a system of the form

$$\prod_{i=1}^{s}(A + \rho_i I)x = y \tag{7.47}$$

has been used by Sweet [13] in connection with the cyclic reduction procedure.

7.7 Rational Iteration

Since A is SPD, the solution $u(t)$ of (7.3) converges to the steady state solution $\bar{u} = A^{-1} b$ as $t \to \infty$. Moreover, we can regard the time dependent schemes considered in the previous section as iterative procedures for solving the linear system (7.1). Thus, corresponding to the backward difference method and the Crank-Nicolsen method, respectively, we have the iterative methods

$$(A + \rho I)u^{(n+1)} = \rho u^{(n)} + b \tag{7.48}$$

and

$$(A + \rho I)u^{(n+1)} = -(A - \rho I)u^{(n)} + 2b . \tag{7.49}$$

These two "rational" iterative methods correspond to the matrix splittings

$$A = (A + \rho I) - \rho I \tag{7.50}$$

and

$$A = \frac{1}{2}(A + \rho I) - [-\frac{1}{2}(A - \rho I)] \tag{7.51}$$

respectively.

One can also consider non-stationary iterative methods where ρ varies. Thus, for example, one could apply (7.48), first with ρ_1 and then with ρ_2 obtaining

$$(A + \rho_1 I)u^{(n+1)} = \rho_1 u^{(n)} + b \tag{7.52}$$

and

$$(A + \rho_2 I)u^{(n+2)} = \rho_2 u^{(n+1)} + b \,. \tag{7.53}$$

From this, it follows that

$$u^{(n+2)} = (A + \rho_2 I)^{-1}(A + \rho_1 I)^{-1}(\rho_1\rho_2 u^{(n)} + \rho_2 b) + (A + \rho_2 I)^{-1}b \tag{7.54}$$

and

$$\begin{aligned} u^{(n+2)} - \bar{u} &= (A + \rho_2 I)^{-1}(A + \rho_1 I)^{-1}\rho_1\rho_2(u^{(n)} - \bar{u}) \\ &= R(A)(u^{(n)} - \bar{u}) \end{aligned} \tag{7.55}$$

where $R(x)$ is the rational function

$$R(x) = \frac{\rho_1\rho_2}{(x + \rho_1)(x + \rho_2)} \,. \tag{7.56}$$

Since $R(x)$ is a rational function we refer to the above procedure as "rational iteration." In this case of polynomial acceleration, $R(x)$ would be a polynomial.

At this point we note that the ordinary extrapolated Richardson's method can be derived by applying the *forward difference method* to (7.3). Thus we have

$$\frac{u(t + \Delta t) - u(t)}{\Delta t} = -Au(t) + b \tag{7.57}$$

or

$$u(t + \Delta t) = (I - (\Delta t)A)u(t) + b\Delta t \,. \tag{7.58}$$

This corresponds to the extrapolated Richardson's method defined by

$$\begin{aligned} u^{(n+1)} &= (I - \gamma A)u^{(n)} + \gamma b \\ &= G_{[\gamma]}u^{(n)} + b_{[\gamma]} \end{aligned} \tag{7.59}$$

where $\gamma = \Delta t$ is the *extrapolation factor*. If the eigenvalues ν of A lie in the range $0 < m(A) \le \nu \le M(A)$ then the optimum extrapolation factor $\gamma*$ is given by

$$\gamma^* = \frac{2}{M(A) + m(A)} \qquad (7.60)$$

and the corresponding spectral radius of $G_{[\gamma*]}$ is

$$S\left(G_{[\gamma*]}\right) = \frac{M(A) - m(A)}{M(A) + m(A)} = \frac{K(A) - 1}{K(A) + 1} \qquad (7.61)$$

where $K(A) = K(A)/m(A)$ is the condition number of A. The number of iterations needed for convergence using the extrapolated Richardson's method is asymptotically proportional to $K(A)$. If conjugate gradient acceleration is used the number of iterations is asymptotically proportional to $K(A)^{1/2}$.

In Figure 7.2 we plot the eigenvalues of the extrapolated Richardson's method in (7.59), ($\lambda_1 = 1 - \gamma\nu$), the eigenvalues of the iteration method of (7.48), ($\lambda_2 = \rho/(\rho + \nu)$) and the eigenvalues of the iterative method of (7.49), ($\lambda_3 = (1 - \nu/\rho)/(1 + \nu/\rho)$). It should be noted that λ_1 vanishes for $\nu = \gamma^{-1}$. If γ is very large, $|\lambda_1|$ is greater than one for large ν. On the other hand, λ_3 vanishes for $\nu = \rho$ but $|\lambda_3| \le 1$ for all ν in the range $m(A) \le \nu \le M(A)$. It should also be noted that λ_2 is a positive monotone decreasing function of ν for $\nu \ge 0$ and that $\lambda_2(0) = 1$.

Let us now consider the iterative method defined by s iterations of (7.49) with variable ρ. The eigenvalues γ of the method are given by

$$\gamma = \frac{\rho_1 - \nu}{\rho_2 + \nu} \frac{\rho_2 - \nu}{\rho_2 + \nu} \cdots \frac{\rho_s - \nu}{\rho_s + \nu} .$$

The rational function given above is the same as that frequently used in the analysis of the Peaceman-Rachford [12] alternating direction implicit scheme; see e.g., Birkhoff, Varga and Young [2]. It can be shown that, for a suitable choice of the $\{\rho_i\}$, the number of iterations required for convergence is asymptotically proportional to $\log K(A)$.

Let us consider the case where the linear system is derived from the standard five-point finite difference representation of the Poisson equation $u_{xx} + u_{yy} = f(x, y)$ in the unit square $0 \le x \le 1$, $0 \le y \le 1$ with Dirichlet boundary conditions. In this case the eigenvalues ν of A lie in the range

$$m(A) = 8 \sin^2 \frac{\pi h}{2} \le \nu \le 8 \cos^2 \frac{\pi h}{2} = M(A) \qquad (7.62)$$

where h is the mesh size. The condition number $K(A)$ of A is given by

$$K(A) = \frac{M(A)}{m(A)} = \cot^2 \frac{\pi h}{2} \approx \frac{4}{\pi^2 h^2} = O(h^{-2}) . \qquad (7.63)$$

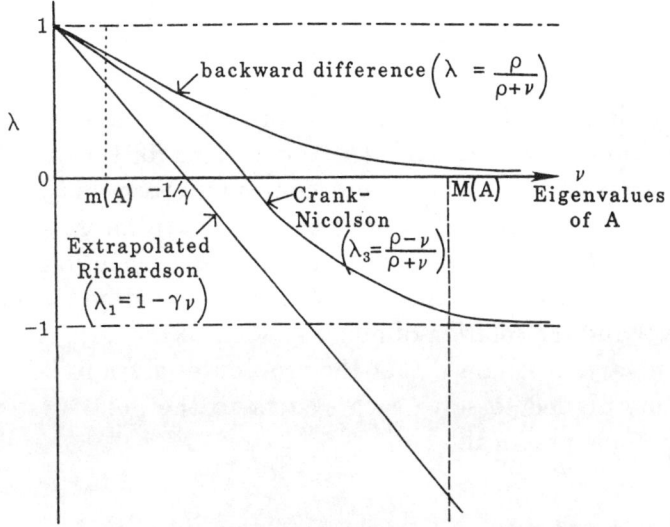

Figure 7.2: Eigenvalues of Iterative Procedures.

Thus using the extrapolated Richardson's method the number of iterations is $O(h^{-2})$, whereas if conjugate gradient acceleration is used the number of iterations is $O(h^{-1})$. Using the Peaceman-Rachford scheme with good parameters, the numbers of iterations is $O(\log h^{-1})$.

Evidently, rational iteration has many attractive properties as compared with polynomial acceleration. Unfortunately there is one serious drawback, namely, the amount of work needed to carry out each iteration. Thus if ρ is very small, the amount of work required to solve a system of the form

$$(A + \rho I)x = y \tag{7.64}$$

for x, given y, may be comparable with that needed to solve the original system (7.1).

As an example, consider the use of a single value of ρ. It can be shown, see, *e.g.*, Birkhoff, Varga and Young [2], that the optimum single value of ρ is given by

$$\rho^* = \sqrt{M(A)m(A)} . \tag{7.65}$$

For the model problem $\rho^* = 4 \sin \pi h = O(h)$. It can be shown that, for the model problem defined above, if conjugate gradient acceleration is applied, the number of iterations is $O(h^{-1/2})$ instead of $O(h^{-1})$ as with the Richardson's method with conjugate gradient acceleration. However, the number of iterations needed to solve[1] each system of the form

$$(A + \rho^* I)x = y \tag{7.66}$$

for x, given y, is also $O(h^{-1/2})$. Thus with $O(h^{-1/2})$ systems each taking $O(h^{-1/2})$ iterations we again have $O(h^{-1})$ iterations for the overall process.

We are now investigating several procedures for overcoming this difficulty. One possibility for solving the system $(A + \rho I)x = y$ for x, given y, would be to do so for a relatively large value of ρ and then let ρ decrease using some kind of continuation method. Another procedure would be a sort of nesting procedure involving the solution of many systems of the form $(A + \rho I)x = y$ where ρ and y vary. It is hoped that the procedures given in Section 7.4 will reduce the time needed to solve each system to the point that the overall scheme will become practical.

Acknowledgments

This work was supported in part by the Department of Energy, under Grant DE-AS05-81ER10954, and the National Science Foundation, under Grant MCS-8214731, with the University of Texas at Austin. Some of the work was done during the fall semester of 1987 while the author was visiting the Oak Ridge National Laboratory and the University of Tennessee under the Special Year on Numerical Linear Algebra.

References

[1] Adams, L. M., *Additive M-Step Preconditioners*, Technical Report 85-6, Department of Applied Mathematics, University of Washington, Seattle, Washington, 1985.

[2] Birkhoff, G., R. S. Varga and D. M. Young, "Alternating Direction Implicit Methods," *Advances in Computers*, 3, 189–273, 1962.

[3] Dubois, P., A. Greenbaum and G. Rodrique, "Approximating the Inverse of a Matrix for Use in Iterative Algorithms on Vector Processors," *Computing*, 22, 257–268, 1978.

[4] Frommer, A., and G. Mayer, "Convergence of Relaxed Parallel Multisplitting Methods," unpublished manuscript, 1988.

---

[1] Here conjugate gradient acceleration would be applied to Richardson's method. The formulas are given in (7.27).

[5] Hageman, L. A., and D. M. Young, *Applied Iterative Methods,* Academic Press, New York, 1981.

[6] Hestenes, M. R., and E. L. Stiefel, "Methods of Conjugate Gradients for Solving Linear Systems," *J. Res. Nat. Bur. Standards,* 49, 409–436, 1952.

[7] Johnson, O., C. Micchelli and G. Paul, "Polynomial Preconditioners for Conjugate Gradient Calculations," *SIAM J. Numer. Anal.,* 20, 362–376, 1983.

[8] Kincaid, D. R., and D. M. Young, *A Review of the ITPACK Project,* Report CNA-217, Center for Numerical Analysis, The University of Texas, Austin, Texas, 1988.

[9] O'Leary, D. P., "The Block Conjugate Gradient Algorithm and Related Methods," *Linear Algebra Appl.,* 29, 293–322, 1980.

[10] O'Leary, D. P., "Parallel Implementation of the Block Conjugate Gradient Algorithm," *Parallel Computing,* 5, 127–139, 1987.

[11] O'Leary, D. P., and R. E. White, "Multi-splittings of Matrices and Parallel Solution of Linear Systems," *SIAM J. Alg. Disc. Meth.,* 6, 630–640, 1985.

[12] Peaceman, D. W., and H. H. Rachford, Jr., "The Numerical Solution of Parabolic and Elliptic Differential Equations," *J. SIAM,* 3, 28–41, 1955.

[13] Sweet, R., "A Parallel and Vector Variant of the Cyclic Reduction Algorithm," *SIAM J. of Sci. Stat. Comp.,* 9, 761–765, (1988).

Chapter 8

Pipelined Successive Overrelaxation

John P. Bonomo* and Wayne R. Dyksen*

8.1   Introduction

Much research effort has been expended toward developing both parallel and vector algorithms for solving the linear equations arising from partial differential equations (PDEs). We present a new iterative method called the pipeline successive overrelaxation (PiSOR) method. PiSOR is the result of our attempt to develop an algorithm which retains the convergence properties of synchronous methods but allows control over the communication between the processors. We give a description of the SOR method in Section 8.2. In Section 8.3 we give a brief survey of previous work on the parallelization of iterative methods. In Section 8.4 we describe the PiSOR method for multiprocessors. The experimental results are presented in Section 8.5. Our conclusions and details of our current work are presented in Section 8.6.

8.2   The SOR Method

Consider the general self-adjoint elliptic PDE on a rectangular domain Ω of the form

$$
\begin{aligned}
-(pu_x)_x - (qu_y)_y + ru &= f(x,y) & (x,y) \in \Omega, \\
u &= g(x,y) & (x,y) \in \partial\Omega
\end{aligned}
\tag{8.1}
$$

---

*Computer Science Department, Purdue University.

where p, q and r are all functions of x and y, and p and q are strictly positive. For simplicity, we assume Dirichlet boundary conditions. We discretize (8.1) by placing an $N + 2$ by $N + 2$ mesh over the domain and by using symmetric finite difference approximations to obtain a set of linear equations

$$Au = f \qquad (8.2)$$

where A is a square matrix of order N^2. The approximate values of u associated with the N^2 interior grid points make up the vector $\mathbf{u}$; we refer to each vector component as a grid component. These grid components are numbered row-wise u_i, $i = 1, \ldots, N^2$, so that grid point (i,j) is associated with $u_{(j-1)N+i}$. Since each grid component is related to its four nearest neighbors, matrix A has bandwidth N.

One standard way to solve (8.2) is the successive overrelaxation (SOR) method. If we let $A = D - E - F$, where D is made up of the diagonal elements of A, and E and F are respectively the strictly lower and upper triangular parts of A, then the SOR method is given by

$$(D - \omega E)u^{(k+1)} = ((1 - \omega)D + \omega F)u^{(k)} + \omega f$$

where ω is the relaxation factor and $0 < \omega < 2$. When $\omega = 1$ the method is referred to as the Gauss-Seidel method. Since we are dealing with a self-adjoint equation, the use of symmetric finite differences insures that A is symmetric positive definite, which in turn insures that the SOR method converges for any initial $\mathbf{u}^{(0)}$ (Varga [13]). Furthermore Young [14] has devised methods to determine the optimal value of the relaxation parameter ω. Using this optimal value of ω increases the rate of convergence of the SOR method by an order of magnitude over that of the Gauss-Seidel method.

8.3 Previous Work

The classical Jacobi algorithm is considered ideal for parallelization since the update of any matrix element in each pass can be done independently of all the other elements. Unfortunately, Jacobi schemes – both serial and parallel – are not attractive due to very slow convergence rates. More useful results have been obtained through parallel modifications of classical Gauss-Seidel and SOR methods. A general survey of these methods is given in Ortega and Voigt [12].

Among the earliest types of modifications are the so called asynchronous methods. Typically in these methods each processor updates a set of grid components, and all processors run simultaneously with minimal communication between them (see Baudet [3] and Kung [10]). No attempt is made to synchronize each iterative sweep. This method avoids two problems which are inherent in any algorithm that attempts to synchronize sweeps. First, extra computational work must be performed by each processor at the end of a sweep to verify when it can start the next sweep. Second, a processor may waste time while waiting for all other processers to finish a sweep. A drawback of asynchronous methods is that analysis of algorithms and proofs of convergence are difficult.

Another major class of parallel iterative methods are multicolor methods. The simplest of these synchronous methods is the Red-Black method in which two colors are assigned to the grid components in a checkerboard manner. All grid components of one color can be updated in a Jacobi-like sweep in odd numbered passes, while those of the second color are updated in even numbered passes. Work has been performed on this scheme by Ericksen [7], Barlow and Evans [2], and Evans [8]. When grid components depend on their diagonal neighbors, such as when the PDE involves cross derivatives or when higher order discretizations are used, orderings involving more than two colors are required. These multicolored orderings are examined by Hackbush [9], Adams and Ortega [1], and O'Leary [11]. While these methods have more well defined convergence properties than asynchronous methods, they are saddled with the overhead costs of keeping the processors synchronized.

8.4 The Pipeline SOR Method

We now present a parallel version of the SOR method called the pipelined successive overrelaxation (PiSOR) method. Let p be the number of processors available denoted by $PE_1, PE_2, \ldots, PE_p$. The basic idea of PiSOR is to perform multiple iterations in parallel on a grid, while spacing the processors so that they do not overlap with each other. Each processor performs one iteration in its entirety, where each iteration proceeds by first updating grid component u_1, then element $u_2, u_3, \ldots, u_{N^2}$. PE_1 performs iterations 1, $p + 1, 2p + 1, \ldots$; PE_2 performs iterations $2, p + 2, 2p + 2, \ldots$ etc. Iterations follow one another in a pipelined fashion; i.e., after iteration 1 has proceeded for a certain amount of time, iteration 2 starts and both iterations update different grid components simultaneously. At the appropriate time, iteration 3 begins, and three updates proceed in parallel through the vector u.

In order to apply traditional error analysis and convergence rate proofs for SOR, we must ensure that no iteration overtakes the one ahead of it. To this end, we introduce a set of synchronization flags assigned to various grid components. We define δ to be the *pipeline spacing* which specifies the spacing of the synchronization flags. Each processor updates δ grid components at a time, starting with u_1. Before updating the grid components $u_i, u_{i+1}, \ldots u_{i+\delta-1}$, a processor must be sure that the previous SOR iteration has updated all the grid components that are needed for its update, specifically u_{i-N} through $u_{i+\delta+N-1}$. In order to ensure this, the processor checks the synchronization flag associated with $u_{i+\delta+N-1}$. If it indicates that $u_{i+\delta+N-1}$ has been updated by the previous iteration, then the processor can continue the current iteration by updating u_i through $u_{i+\delta-1}$ (the fact that $u_{i+\delta+N-1}$ has been updated implies that all grid components u_{i-N} through $u_{i+\delta+N-2}$ have also been updated by the previous iteration). Before proceeding to the next set of δ grid components, the processor updates the synchronization flag associated with $u_{i+\delta-1}$ to signal the next iteration. Note that each iteration leads its successor by $(N + \delta)$ grid components.

In general, when a synchronization flag has value k, this indicates that iterations $1, 2, \ldots k - 1$ have updated the grid components corresponding to the synchronization flag, and that iteration k may proceed in using these grid components to perform its updates. Initially, all synchronization flags are set to 1, allowing PE_1 to perform iteration 1 unhindered.

For given values of N and δ, synchronization flags, s_i, are associated with grid components $u_{\delta+N}, u_{2\delta+N}, \ldots, u_{M\delta+N}$, where M is the smallest integer such that $M\delta + N \geq N^2$, i.e.,

$$M = \left\lceil \frac{N^2 - N}{\delta} \right\rceil .$$

Note that once a PE begins to update the last N grid components (corresponding to the last row of the grid), it does not need to check any synchronization flags, since it no longer needs any grid components updated by the previous iteration. Synchronization check points, c_i, are associated with grid components $u_1, u_{\delta+1}, \ldots, u_{(M-1)\delta+1}$. At each check point c_i, each processor must check synchronization flags s_i. Figure 8.1 shows the interior grid points and the placement of the synchronization flags for $N = 6$ with $\delta = 4$ and 5. The black circles indicate the location of the synchronization check points and the square boxes indicate the location of the synchronization flags. In Figure 8.1a, the first synchronization flag is associated with grid component $u_{\delta+N} = u_{10}$ and every subsequent fourth grid component. In Figure 8.1b, the first flag is associated with grid component u_{11} and every fifth grid component from then on. In both examples, the first check point appears at grid component u_1 and subsequent ones appear every δ grid components thereafter.

An illustration of the PiSOR algorithm is given in Figure 8.2 for the case $N = 6, \delta = 6$. In Figure 8.2a, all processors but PE_1 are waiting on s_1 while PE_1 begins updating grid components. Before it updates u_7, it checks s_2's value. Since that value is 1, PE_1 proceeds to update the next 6 ($= \delta$) grid

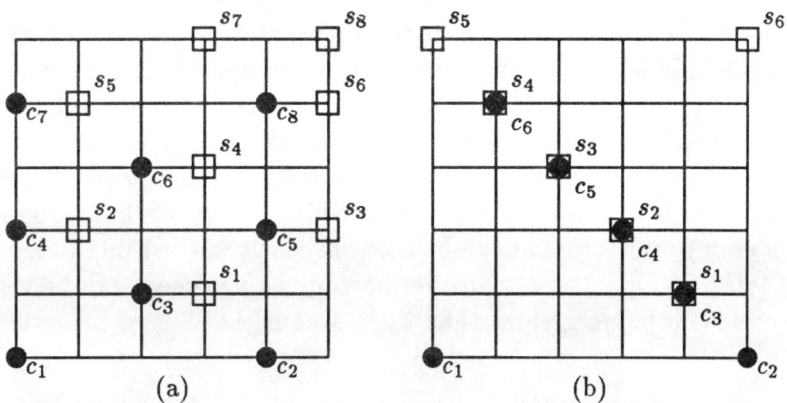

Figure 8.1: PiSOR synchronization flag placement with (a) $N = 6, \delta = 4$ and (b) $N = 6, \delta = 5$. Circles indicate flag check points and squares indicate synchronization flags. Only interior grid points are shown.

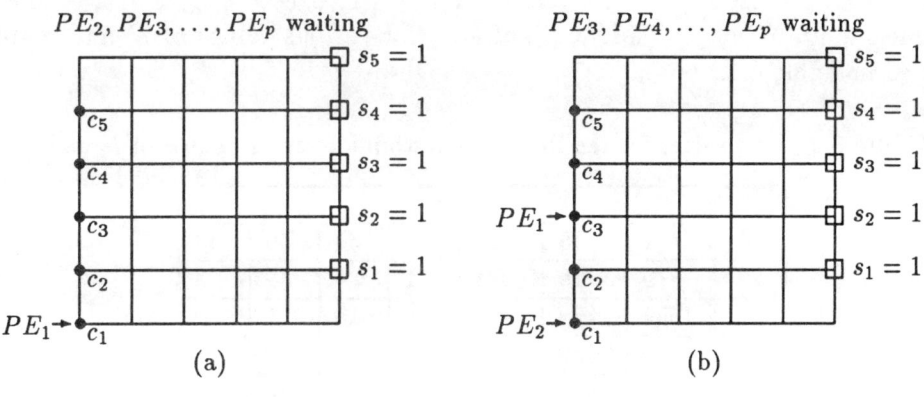

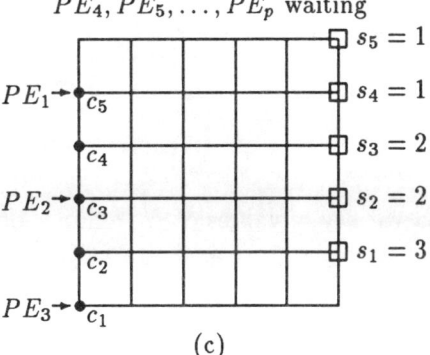

Figure 8.2: Sample PiSOR iterations with $N = 6, \delta = 6$. Only interior grid points are shown. Note that in practice, the use of more than three processors for these values of N and δ would be inefficient.

components. After it updates u_{12} it changes s_1's values to 2; since $s_3 = 1$, PE_1 proceeds to update u_{13} (Figure 8.2b). Since $s_1 = 2$, PE_2 starts the second iteration. Figure 8.2c shows one more step in this process. After PE_1 finishes iteration 1 and updates s_5, it waits on synchronization flag s_1 until its value is $p+1$, in which case it starts performing iteration $p+1$, and so on.

The application of pipeline techniques to a serial iterative method introduces certain constraints in the resulting parallel method. Since iterations are separated from each other by at least $(N + \delta)$ grid components, at most $[(N^2 - 1)/(N + \delta)] + 1$ iterations can be pipelined at once. Thus, the maximum number of processors that can be used efficiently in the PiSOR algorithm is $[(N^2 - 1)/(N + 1)] + 1 \equiv N$. Furthermore, given p processors, the value of δ is constrained to be no greater than $\delta_{\max}$, where

$$\delta_{\max} = \left[\frac{N^2 - 1}{p - 1}\right] - N . \tag{8.3}$$

Using a value for $\delta > \delta_{\max}$ will result in inefficient use of some processors throughout the execution of the algorithm, since the spacing between processors will be too big to allow all processors to update grid components simultaneously. Note that in the example in Figure 8.2, if more than three processors are used, then the value of $\delta = 6$ will exceed $\delta_{\max}$ and a smaller value should be used. The values of $\delta_{\max}$ for various values of N and p are listed in Table 8.1.

Table 8.1: $\delta_{\max}$ values for the PiSOR algorithm for various values of N and p.

	p						
N	2	5	10	20	30	40	50
20	379	79	24	1	–	–	–
40	1559	359	137	44	15	1	–
60	3539	839	339	129	64	32	13
80	6319	1519	631	256	140	84	50
100	9899	2399	1011	426	244	156	104

8.5   Experimental Results

The PiSOR algorithm was applied for a variety of values of N, δ and p on the following elliptic problem:

$$-u_{xx} - u_{yy} = -4 \qquad \text{in } \Omega = [0,1] \times [0,1]$$
$$u = x^2 + y^2 + y \qquad \text{on } \partial\Omega,$$

for which the solution is $u = x^2 + y^2 + y$. We use the optimal value for the relaxation factor

$$\omega = \frac{2}{1 + \sin(\pi/(N+1))} \, .$$

The experiments were run on a Sequent Symmetry multiprocessor (further details on the machine can be found in Bonomo and Dyksen [6]). The number of available processors was 27. Tables 8.2 through 8.6 show the experimental results for $N^2 = 400, 1600, 3600, 6400$ and 10000, $\delta = 1, 2, 5, 10, 20, 50$ and 100, and $p = 1, 2, 4, 8, 11, 16, 20, 24$ and 27. Table 8.7 shows the execution times and iteration counts for sequential SOR. All times represent only the solution time; they do not include startup and discretization time.

Speedups for five values of N with $\delta = 1, 10$ and 100 are graphed in Figure 8.3. We use the standard definition of speedup, $S(p)$, namely

$$S(p) = \frac{\text{Sequential SOR time}}{\text{PiSOR time using } p \text{ processors}} \, .$$

As expected, when one processor is used, the speedup is < 1 due to the overhead involving parallel processing system calls. Note also that as the number of δ becomes greater than δ_{max} ($= [(N^2 - 1)/(p - 1)] - N$), the speedups decrease significantly as expected. This effect is most clearly seen with $\delta = 100$.

The efficiency of the PiSOR algorithm is shown in Figure 8.4 for $N = 40$, $60, 80$ and 100 where the efficiency is defined by $E(p) = S(p)/p$. For $N = 100$ and a suitable choice of δ, efficiency levels over 0.8 can be achieved even when using 20 processors. When $p \leq 4$, efficiency levels over 0.9 are obtained for a wide range of δ values. Note again that once δ_{max} is reached, the efficiency of the algorithm diminishes rapidly.

Table 8.2: PiSOR Execution times (seconds) for $N^2 = 400$.

δ	Number of Processors								
	1	2	4	8	11	16	20	24	27
1	1.660	.903	.470	.267	.207	.167	.170	.173	.177
2	1.617	.867	.460	.257	.203	.160	.163	.173	.170
5	1.590	.857	.450	.250	.197	.163	.180	.183	.180
10	1.580	.857	.440	.253	.200	.190	.200	.213	.210
20	1.580	.843	.437	.250	.223	.233	.250	.257	.263
50	1.573	.833	.433	.343	.353	.383	.400	.417	.437
100	1.580	.823	.517	.553	.573	.610	.643	.673	.703

Table 8.3: PiSOR Execution times (seconds) for $N^2 = 1600$.

δ	Number of Processors								
	1	2	4	8	11	16	20	24	27
1	13.180	6.950	3.627	1.907	1.463	1.070	0.923	0.803	0.750
2	12.747	6.767	3.460	1.820	1.393	1.027	0.873	0.773	0.713
5	12.527	6.603	3.353	1.773	1.363	1.000	0.853	0.757	0.693
10	12.450	6.517	3.327	1.760	1.350	0.997	0.850	0.750	0.690
20	12.440	6.523	3.327	1.760	1.353	1.000	0.850	0.753	0.727
50	12.397	6.403	3.287	1.750	1.350	1.017	0.957	0.980	1.000
100	12.457	6.430	3.283	1.757	1.367	1.380	1.407	1.447	1.487

Table 8.4: PiSOR Execution times (seconds) for $N^2 = 3600$.

δ	Number of Processors								
	1	2	4	8	11	16	20	24	27
1	42.230	21.697	11.247	5.947	4.457	3.263	2.683	2.340	2.140
2	40.747	21.283	10.883	5.677	4.257	3.110	2.577	2.233	2.030
5	40.173	20.997	10.583	5.517	4.157	3.047	2.520	2.170	1.977
10	39.683	20.823	10.473	5.460	4.120	3.017	2.487	2.150	1.953
20	39.500	20.747	10.437	5.430	4.117	3.013	2.470	2.157	1.960
50	39.737	20.360	10.420	5.440	4.083	2.997	2.467	2.177	2.000
100	39.757	20.530	10.450	5.437	4.063	2.973	2.527	2.350	2.380

Table 8.5: PiSOR Execution times (seconds) for $N^2 = 6400$.

δ	Number of Processors								
	1	2	4	8	11	16	20	24	27
1	107.778	54.644	27.972	14.661	10.944	7.883	6.417	5.517	5.083
2	104.267	53.144	27.028	14.139	10.467	7.472	6.122	5.278	4.828
5	102.333	52.158	26.700	13.625	10.208	7.308	5.958	5.125	4.683
10	101.583	52.350	26.467	13.525	10.083	7.242	5.917	5.092	4.667
20	101.300	51.933	26.392	13.475	10.067	7.192	5.892	5.092	4.667
50	100.950	51.583	26.333	13.442	10.025	7.175	5.908	5.092	4.675
100	100.992	51.758	26.175	13.333	10.025	7.158	5.950	5.150	4.717

Table 8.6: PiSOR Execution times (seconds) for $N^2 = 10000$.

δ	Number of Processors								
	1	2	4	8	11	16	20	24	27
1	196.417	99.000	50.650	26.317	19.533	13.900	11.350	9.828	8.906
2	190.083	95.811	49.106	25.406	18.828	13.328	11.039	9.422	8.550
5	186.217	93.917	47.750	24.825	18.308	12.992	10.667	9.183	8.392
10	184.883	94.092	47.583	24.383	18.075	12.892	10.583	9.108	8.325
20	184.425	93.475	47.467	24.350	18.008	12.842	10.550	9.075	8.292
50	184.108	93.750	47.258	24.283	18.017	12.825	10.533	9.100	8.233
100	183.850	93.175	47.425	24.400	18.100	12.767	10.483	9.083	8.167

Table 8.7: Sequential SOR execution times (seconds) and number of iterations.

N^2	Time	Number of Iterations
400	1.493	66
1600	11.713	131
3600	37.433	184
6400	95.650	262
10000	174.083	306

8.6 Conclusion

The PiSOR algorithm presented in this paper is a new efficient method for parallelizing SOR. The algorithm sustains a high level of efficiency in processor usage while also maintaining the convergence properties of the original serial algorithm. While the experiment presented in this paper was motivated by the discretization of an elliptic PDE, the PiSOR algorithm can be applied to any linear system which can be solved using SOR. This pipeline technique is readily applicable to SOR methods which vary the value of ω. Moreover, the methods described in this paper can be used to parallelize any iterative method which is based on a splitting of the linear system matrix A (Bonomo and Dyksen [6]).

The pipelining technique described here can be readily applied to other numerical methods. The pipeline spacing parameter has been used in a modified form to generalize a parallel algorithm created by Patel and Jordan. The original algorithm assigns to each processor the task of updating one row of grid components. Since SOR re-uses updated values as soon as they are available, each processor must wait for the previous processor's iterative updates before it can begin updating the components in its row. Synchronization between the processors is controlled by flags assigned to each grid component. Our generalization of the Patel-Jordan algorithm involves the introduction of a spacing parameter which specifies how many grid components each processor updates prior to signaling the next processors. When this spacing parameter is set equal to one, the original Patel-Jordan algorithm is recovered. Again, the spacing parameter serves as a fine-tuning mechanism to achieve maximum efficiency. Our results show that the efficiency is improved for spacings larger than one (Bonomo and Dyksen [5]).

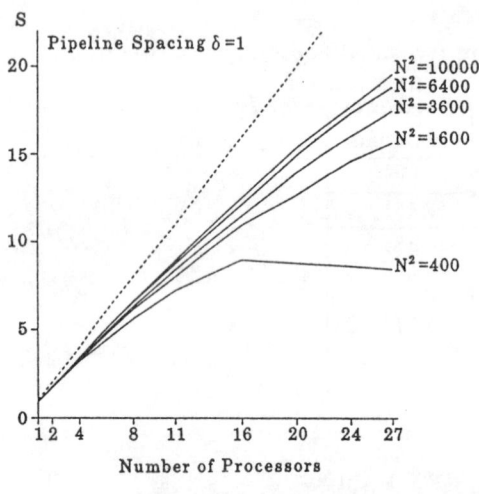

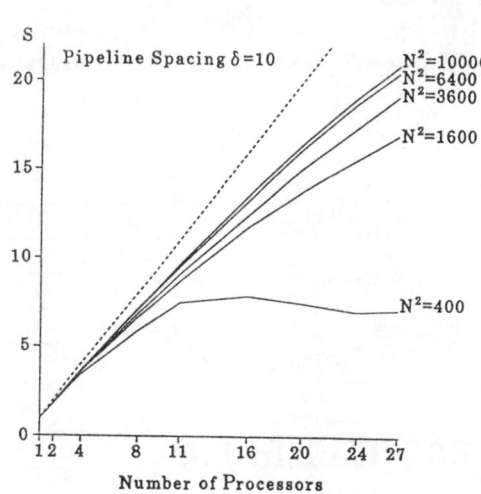

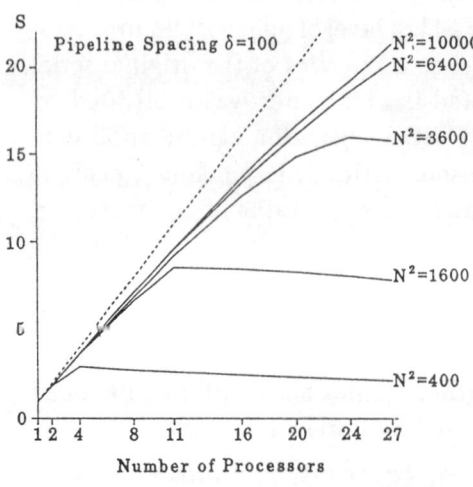

Figure 8.3: Speedups, S, for pipeline spacing $\delta = 1$, 10, and 100. The dotted lines represent optional speedup. Computed speedup increases monotonecally with N as indicated.

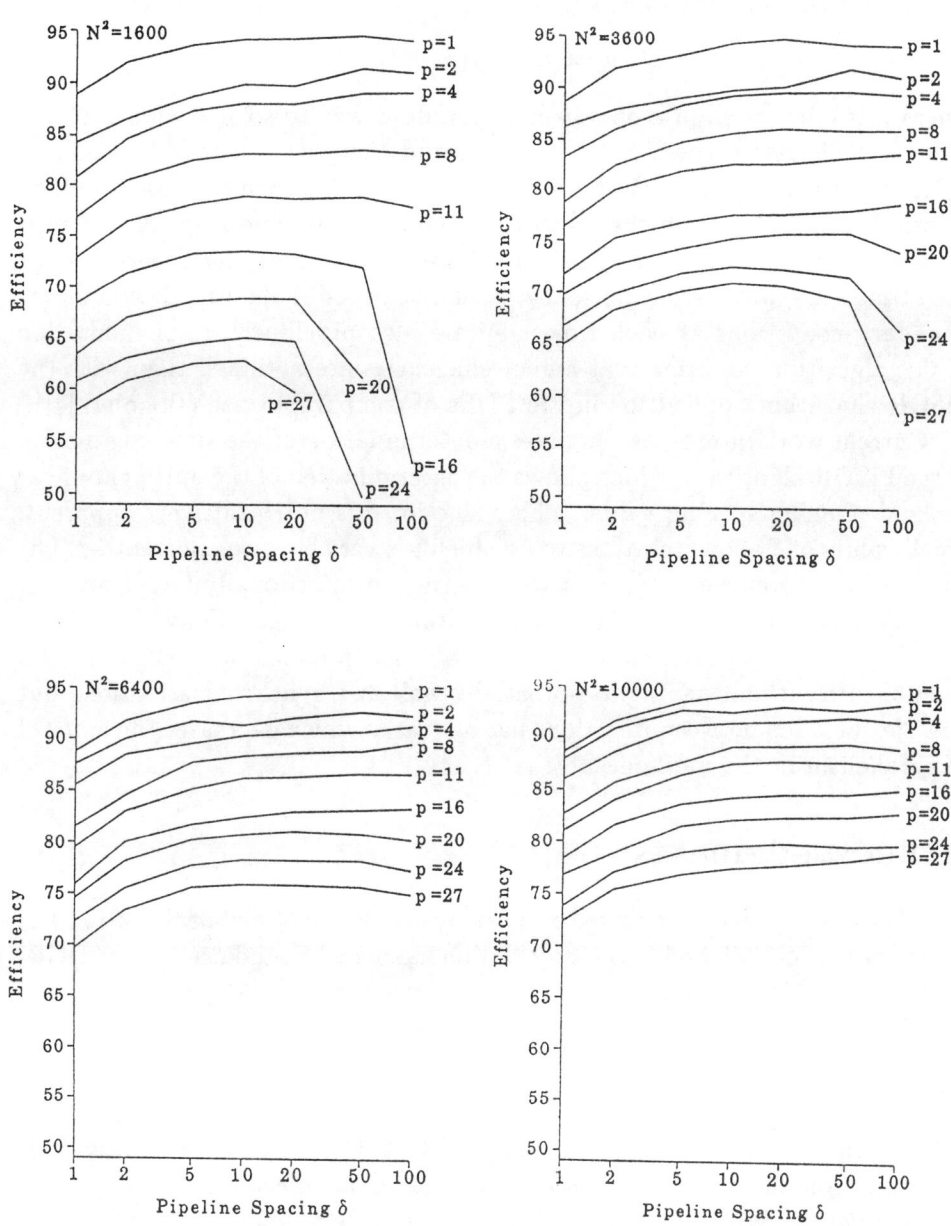

Figure 8.4: Efficiencies for $N^2 = 1600$, 3600, 6400 and 10000.

Pipeline techniques can also be applied to the solution of time-dependent parabolic PDEs of the form

$$u_t = Lu + f(x, y, t),$$

where L is a linear elliptic operator. A standard way to solve such a problem is to first discretize the time dimension, and then solve a modified elliptic PDE at each time step. The boundary conditions for each of these modified elliptic PDEs depends on the solution at the previous time step. A first pass at pipelining this method would be to simply apply the PiSOR algorithm at each time step, but better efficiency can be achieved if the calculation of the boundary conditions at each time step are also pipelined. Implementation of this algorithm confirms that higher efficiencies are obtained than with the PiSOR algorithm applied to elliptic PDEs of comparable size (Bonomo [4]).

Current work in progress includes a determination of the time complexity of the PiSOR algorithm, which allows the determination of the optimal or near optimal pipeline spacing value. This value for δ is not intuitively apparent. Small values of δ allow iterations to be pipelined very close to one another, but consequently force each iteration to perform a proportionally larger amount of work checking and updating the synchronization flags. Conversely, large values of δ allow each iteration process a larger number of contiguous grid components without having to potentially wait at a synchronization flag, but now the distance between iterations has increased which decreases the amount of parallelism in the method.

Acknowledgments

This research was supported in part by Air Force Office of Scientific Research grant 84-0385 and in part by National Science Foundation grant DCR-8602385.

References

[1] Adams, L. M., and J. M. Ortega, "A Multi-Color SOR Method for Parallel Computation," in *Proceedings of the 1982 International Conference on Parallel Processing*, K. E. Batcher, *et al.* (eds.), 53–56, 1982.

[2] Barlow, R. H., and D. J. Evans, "Parallel Algorithms for the Iteative Solution of Linear Systems," *Computer Journal*, 25, 1, 56–60, 1982.

[3] Baudet, G., "Asynchronous Iterative Methods for Multiprocessors," *Journal of the ACM*, 25, 2, 226–244, 1978.

[4] Bonomo, J. P., *Parallel Iterative Techniques for the Solution of Elliptic Partial Differential Equations,* Doctoral Thesis, Purdue University, 1989.

[5] Bonomo, J. P., and W. R. Dyksen, *A Generalization of the Patel-Jordan Parallel Iterative Technique,* Technical Report 811, Purdue University, 1988.

[6] Bonomo, J. P., and W. R. Dyksen, "Pipeline Iterative Methods for Shared Memory Machines," *Parallel Computing,* to appear, 1989.

[7] Ericksen, J., *Iterative and Direct Methods for Solving Poisson's Equation and Their Adaptability to ILLIAC IV,* Center for Advanced Computation Document 60, University of Illinois, Urbana–Champaign, 1972.

[8] Evans, D. J., "Parallel S.O.R. Iterative Methods," *Parallel Computing,* 1, 3–18, 1984.

[9] Hackbush, W., "On the Multigrid Method Applied to Difference Equations," *Computing,* 20, 291–306, 1978.

[10] Kung, H., "Synchronized and Asynchronous Parallel Algorithms for Multi-Processors," in *Algorithms and Complexity,* J. F. Traub (ed.), Academic Press, 153–200, 1976.

[11] O'Leary, D. P., "Ordering Schemes for Parallel Procesing of Certain Mesh Problems," *SIAM Journal of Scientific and Statistical Computing,* 5, 3, 620–632, September 1984.

[12] Ortega, J. M., and R. G. Voigt, "Solution of Parital Differential Equations on Vector and Parallel Computers," *SIAM Review,* 27, 2, 149–240, 1985.

[13] Varga, R. S., *Matrix Iterative Analysis,* Prentice Hall, 1962.

[14] Young, D. M., *Iterative Solution of Large Linear Systems,* Academic Press, 1971.

Chapter 9

Some Parallel Algorithms on the Four Processor Cray X-MP4 Supercomputer

David R. Kincaid and *Thomas C. Oppe**

9.1 Introduction

To gain experience in the development of parallel algorithms, three numerical studies have been conducted using microtasking on the four processor Cray X-MP4 supercomputer. The first two studies concern software for solving large sparse linear systems of algebraic equations by iterative methods. The first study involves the development of an experimental version of the computer package ITPACKV 2C for parallel execution on this computer. The second study deals with the writing of a parallel version of the conjugate gradient method with line Jacobi preconditioning. The third study examines some parallel algorithms for computing the LU–factorization of a dense matrix. In this chapter, we will briefly review each investigation.

9.2 Microtasked ITPACK

ITPACK 2C is a collection of seven iterative algorithms for solving large sparse linear systems of equations: Jacobi Conjugate Gradient (JCG), Jacobi Semi-Iteration (JSI), Successive Overrelaxation (SOR), Symmetric SOR

---

*Center for Numerical Analysis, The University of Texas at Austin, TX.

Conjugate Gradient (SSORCG), Symmetric SOR Semi-Iteration (SSORSI), Reduced System Conjugate Gradient (RSCG), and Reduced System Semi-Iteration (RSSI). Each method can be used in conjunction with either the natural ordering or a red-black ordering applied to the unknowns except for the Reduced System methods which require a red-black ordering. There are a number of reports on this package that should be consulted for further information (see, for example, Kincaid, Respess, Young, and Grimes [19] and Kincaid and Young [20]).

Vectorized versions of the ITPACK 2C package, denoted ITPACKV 2C, were developed for the Cyber 205 and Cray X-MP computers. For information regarding the vectorization techniques used in adapting the package for these two computers, the reader should consult Kincaid and Oppe [15], Kincaid, Oppe and Young [18] and Kincaid, Oppe, Respess and Young [17]. It was found that the JCG and JSI algorithms vectorized well with long vector lengths when the natural ordering was used, and all seven algorithms vectorized well with long vector lengths when red-black ordering was used. In both cases, the algorithms could be expressed in terms of four basic vectorizable operations: dot products, SAXPYs, simple vector operations, and matrix-vector products. (SAXPY denotes a Basic Linear Algebra Subprogram (Lawson, Hanson, Kincaid, and Krogh [21]) for computing $y \leftarrow y + \alpha x$ for vectors x and y and a scalar α.) Thus, it was hoped that these algorithms could be effectively parallelized by partitioning each vectorized operation into disjoint suboperations which could be assigned to different processors, as in Seager [25]. Using this philosophy, a parallel version of ITPACKV 2C was written for the Cray X-MP4 computer using Cray microtasking tools. Since the vectorizable DO loops were partitioned, each CPU updates separate quantities so data protection was unnecessary. However, bank conflicts are still possible since each CPU executes in a nondeterministic order. The goals were to minimize the number of synchronizations (i.e., maximize the amount of work between synchronizations) and to load balance the "physical" CPUs. As much independent work as possible was placed between synchronization points. Synchronization points were forced if the result of a dot product was needed and also prior to a matrix-vector product. All vectorizable DO loops were executed in parallel mode, and scalar code was executed in uniprocessor mode.

Some observations can be drawn from the microtasking of ITPACK. First, microtasked code must be separated into its own subroutines, so massive changes to the vector ITPACKV code were required, but not as massive as the standard Cray multi-tasking tools would have required. These "artificial" subroutines were constructed to perform the maximum possible work between synchronization points. It should be noted that ITPACK was not intended

for multiprocessing and that vectorizable work is often interrupted by parameter estimations, eigenvalue computation, etc. Second, the microtasked code
was completely debugged on our local uniprocessor Cyber 170/750 computer
since microtasking directives introduce nothing nonportable. In general, true
debugging of a parallel code needs to be done on a multiprocessor. Finally,
there are some costs associated with microtasking. A larger number of subroutine calls is incurred. "STACK-based" loading must be used instead of
"STATIC-based." With STATIC, only one copy of each subroutine exists.
With STACK, a copy of the subroutine is made for each call which is slower.
Also, there is some overhead involved in breaking tasks into pieces and possible memory bank conflicts.

The model problem for the numerical experiments in this paper is the
following elliptic partial differential equation with Dirichlet boundary conditions:

$$\begin{cases} u_{xx} + 2u_{yy} = 0 & \text{on } S = [0,1] \times [0,1] \\ u = 1 + xy & \text{on the boundary of } S \end{cases}$$

When the region S is discretized into a square grid of mesh size h and the partial derivatives are approximated by the standard central finite-difference formulas, the result is a sparse linear system of equations $Au = b$ with no more
than five nonzeros u_j in each row. This coefficient matrix becomes extremely
large as the mesh size h decreases but retains the same sparsity pattern—a
large sparse linear system. To simulate a more realistic problem, we do not
take advantage of the fact that the operator has constant coefficients. In
this study, the region is taken to be the unit square, mesh size $h = \frac{1}{100}$, and
normalized residual tolerance for the stopping test is $\epsilon = 5 \times 10^{-6}$. In Tables 9.1 and 9.2, the number of iterations required for convergence and the
timing results in seconds are given for the natural and a red-black ordering
of the unknowns. The number of iterations required for convergence changed
for two of the ITPACK routines in the microtasked version and this is noted
in the tables. The uniprocessor Cray X-MP4 version (STATIC allocation) of
the code uses calls to SECOND for timing while the multiprocessor version
(STACK allocation) uses calls to the real time clock IRTC to measure wall
clock time. In the multiprocessor version, the number of tasks were varied
from one ($t = 1$) to four ($t = 4$). In these tables, the numbers in parentheses
are the timing results divided into the corresponding ones for one task ($t = 1$),
i.e., the "speedup." Moreover, all results are given for runs made in dedicated
machine mode.

From these results, we can draw a number of conclusions. When comparing speedup ratios between the uniprocessor version and the multiprocessor
version, we see that there is an inherent cost involved in microtasking. For

Table 9.1: Microtasked ITPACK: Natural Ordering.

Method	Iter.	Uniprocessor	Number of Tasks			
			t=1	t=2	t=3	t=4
JCG	324	.679	.879	.463	.333	.267
	(328)	(1.295)	(1.000)	(1.898)	(2.640)	(3.292)
JSI	534	.974	.993	.527	.383	.312
		(1.020)	(1.000)	(1.884)	(2.593)	(3.183)
SOR	302	7.654	7.862	7.651	7.589	7.559
		(1.027)	(1.000)	(1.028)	(1.036)	(1.040)
SSORCG	47	2.703	2.771	2.655	2.621	2.609
		(1.025)	(1.000)	(1.044)	(1.057)	(1.062)
SSORSI	55	2.987	3.057	2.955	2.925	2.933
		(1.023)	(1.000)	(1.035)	(1.045)	(1.042)

both orderings, the best speedup per number of tasks was obtained with only two tasks (several methods having 1.8-1.9 speedup on two processors). The JCG method with four tasks was the fastest method using natural ordering but several methods (SOR, RSCG, RSSI) were faster than JCG when using a red-black ordering. In fact, the RSCG method (red-black ordering) was the fastest overall method on four processors.

9.3   Parallel Line Jacobi Preconditioned Conjugate Gradient

In the second investigation, a parallel version of the conjugate gradient method with line Jacobi preconditioning (LJCG) was implemented on a four processor Cray X-MP4 supercomputer using microtasking. This algorithm was tested by solving the model elliptic partial differential equation on the unit square that was described in the previous section but with varying mesh sizes. In this method, the preconditioning step involves the solution of a tridiagonal system $Tz = r$, where T is the tridiagonal part of A. T is factored once before the iteration commences, and a tridiagonal solve is performed on each iteration.

The operations of factoring a tridiagonal matrix and solving a tridiagonal system are, in general, difficult to vectorize. However, in this case, T consists of multiple independent subsystems of equal size. Hence, it is possible to apply the recursive operations used to factor and solve one subsystem across

Table 9.2: Microtasked ITPACK: Red-Black Ordering.

Method	Iter.	Uniprocessor	Number of Tasks			
			t=1	t=2	t=3	t=4
JCG	324	.687	.883	.483	.352	.356
	(328)	(1.285)	(1.000)	(1.828)	(2.509)	(2.480)
JSI	534	.986	1.066	.555	.411	.412
		(1.081)	(1.000)	(1.921)	(2.594)	(2.587)
SOR	314	.504	.591	.329	.247	.269
		(1.173)	(1.000)	(1.796)	(2.393)	(2.197)
SSORCG	709	3.964	4.572	2.473	1.829	1.951
		(1.153)	(1.000)	(1.849)	(2.500)	(2.343)
SSORSI	279	1.102	1.204	.659	.493	.547
		(1.093)	(1.000)	(1.827)	(2.442)	(2.201)
RSCG	164	.285	.317	.179	.136	.114
	(165)	(1.112)	(1.000)	(1.771)	(2.331)	(2.781)
RSSI	301	.427	.431	.245	.190	.222
		(1.009)	(1.000)	(1.759)	(2.268)	(1.941)

all the subsystems in a vectorizable way. (To simulate a more complicated problem, we again did not take advantage of the fact that the subsystems are identical for this model problem.) If the subsystems are "stacked" together, then corresponding elements of all the subsystems are a constant stride apart. Thus, since the Cray X-MP4 can do vector operations involving a constant stride, the factor and solution algorithms can be vectorized by applying each step to corresponding elements of all the subsystems. Also, the line Jacobi preconditioner is easy to parallelize since the subsystems are independent. If p processors are available and there are NSYS independent subsystems, then each processor is assigned approximately NSYS/p subsystems to factor and solve.

To describe the parallel line Jacobi conjugate gradient algorithm, we use the following notation. Given p processors, define the integer vectors IBGN and IEND by

$$IBGN(s) = [(s-1) * NSYS/p] * NSIZE + 1$$
$$IEND(s) = [s * NSYS/p] * NSIZE$$

for $s = 1, 2, \ldots, p$ where NSYS is the number of subsystems, each of size NSIZE. Here, $[\cdot]$ denotes the greatest integer function. Now we introduce the notation $[\boldsymbol{x}]_s$ to denote the vector $(x_{\text{IBGN}(s)}, \ldots, x_{\text{IEND}(s)})^T$. With this notation, a parallel version of the line Jacobi conjugate gradient (LJCG) algorithm can be written as:

For $n = 0, 1, 2, \ldots$ until convergence do

$$
\left[\boldsymbol{r}^{(n)}\right]_s =
\begin{cases}
[\boldsymbol{b}]_s - \left[A\boldsymbol{u}^{(0)}\right]_s & \text{if } n = 0 \\
\left[\boldsymbol{r}^{(n-1)}\right]_s - \alpha_{n-1} \left[\boldsymbol{z}^{(n-1)}\right]_s & \text{if } n > 0
\end{cases}
$$

$$
\left[\boldsymbol{z}^{(n)}\right]_s = \left[T^{-1}\right]_s \left[\boldsymbol{r}^{(n)}\right]_s
$$

$$
\gamma_{n,s} = \left\langle \left[\boldsymbol{z}^{(n)}\right]_s, \left[\boldsymbol{r}^{(n)}\right]_s \right\rangle
$$

—————— sync ————————————————————

$$
\gamma_n = \sum_{s=1}^{p} \gamma_{n,s}
$$

stopping test

$$
\beta_n = \gamma_n / \gamma_{n-1}
$$

—————— fork ————————————————————

$$
\left[\boldsymbol{p}^{(n)}\right]_s = \left[\boldsymbol{z}^{(n)}\right]_s + \beta_n \left[\boldsymbol{p}^{(n-1)}\right]_s
$$

—————— sync ————————————————————

—————— fork ————————————————————

$$
\left[\boldsymbol{z}^{(n)}\right]_s = \left[A\boldsymbol{p}^{(n)}\right]_s
$$

$$
\rho_{n,s} = \left\langle \left[\boldsymbol{p}^{(n)}\right]_s, \left[\boldsymbol{z}^{(n)}\right]_s \right\rangle
$$

—————— sync ————————————————————

$$
\rho_n = \sum_{s=1}^{p} \rho_{n,s}
$$

$$
\alpha_n = \gamma_n / \rho_n
$$

—————— fork ————————————————————

$$
\left[\boldsymbol{u}^{(n+1)}\right]_s = \left[\boldsymbol{u}^{(n)}\right]_s + \alpha_n \left[\boldsymbol{p}^{(n)}\right]_s
$$

The term "sync" indicates a global synchronization of all processors, after which a single processor is computing until a "fork," which indicates multi-processing is resumed for all processors. In this version of the parallel LJCG method, a global synchronization is needed at three places:

- when the results from the first dot product are needed

- before the matrix-vector product $z = Ap$, and

- when the results from the second dot product are needed

The matrix vector product $z = Ap$ is calculated as $[z]_s = [Ap]_s$; that is, processor s computes z_i for rows $i = \text{IBGN}(s)$ to $\text{IEND}(s)$.

Another means of vectorizing the solution of $Tz = r$ is to use a banded approximation to T^{-1}. This makes the preconditioning step $z = \text{band}(T^{-1})r$ a vectorizable matrix-vector multiply. Using a procedure described by Axelsson [1984], it is possible to form the elements of T^{-1} within a band without computing or storing its elements outside the band. Since T is strongly diagonally dominant for the model problem used, T^{-1} is accurately approximated with a tridiagonal matrix. Again, this preconditioner is easy to parallelize since the tridiagonal subsystems are independent.

The mesh sizes used were $h = \frac{1}{16}, \frac{1}{32}, \frac{1}{64}, \frac{1}{128}$, and $\frac{1}{256}$ giving problem sizes of $n = 225, 961, 3969, 16129$, and 65025. The constant vector stride used in this preconditioner is $m = \sqrt{n}$ which is the number of x-grid lines. (To avoid biasing the results by introducing bank conflicts, this number should not be a power of two.) The residual stopping test with $\epsilon = 10^{-6}$ was used. Two versions of line Jacobi were implemented—an exact inverse version using tridiagonal solves and an approximate inverse version using inverses truncated to tridiagonal form. Both versions were implemented with a uniprocessor code and a multiprocessor code using microtasking. Vectorization and parallelism were used whenever possible for both the factorization and solution routines.

We present some representative results from this study showing timings from the uniprocessor and multiprocessor versions of the code for the largest system of size $n = 65025$ corresponding to mesh size $h = \frac{1}{256}$. Table 9.3 gives the timings in seconds for the uniprocessor version and the multiprocessor version of the LJCG method with the exact inverse and the approximate inverse for T. Two timings are reported. The top number is the time to factor T (uniprocessor), the time to obtain the exact inverse of T, or the time to form a tridiagonal approximation to it. The bottom number is the total iteration time for all iterations. The numbers in parentheses give the speedup over one processor.

From the numerical results, it can be seen that the approximate inverse line Jacobi method can achieve greater speedups than the exact inverse line Jacobi method. One possible explanation is that vector lengths for the solution of $Tz = r$ are approximately n/p for the approximate inverse method but only $\sqrt{n}/p$ for the exact inverse method. Since $n \gg \sqrt{n}$, the vector start-up times are negligible for small p in the former case but may be very important in the latter case. Also, the approximate inverse method reads from contiguous locations in memory while the exact inverse method accesses memory

Table 9.3: Uniprocessor and Multiprocessor LJCG.

Method	n	Uniprocessor	Number of Processors			
	(Iter.)		$p = 1$	$p = 2$	$p = 3$	$p = 4$
Exact	65025	.007864	.008400	.004829	.004062	.003517
Inverse	(539)	(1.068)	(1.000)	(1.739)	(2.068)	(2.388)
		8.14272	8.38385	4.67418	3.66933	3.16093
		(1.029)	(1.000)	(1.795)	(2.284)	(2.653)
Approx.	65025	.011068	.011750	.006885	.005068	.004388
Inverse	(539)	(1.062)	(1.000)	(1.707)	(2.319)	(2.678)
		6.87376	6.97032	3.55223	2.62447	2.25768
		(1.014)	(1.000)	(1.962)	(2.656)	(3.087)

locations with a non-unit stride. This method was significantly faster on four processors than it was on one. However, the best speedup per number of processors is given with only two processors. In addition, the cost associated with microtasking this algorithm is negligible. Finally, it can be seen from the speedups in factorization time that speedups are possible with microtasking for tasks with very small granularity.

For additional details on this study, see the report by Oppe and Kincaid [23].

9.4 Parallel *LU*-Factorization Algorithms

Several parallel algorithms for computing the *LU*–factorization of a dense matrix are the subject of the next investigation. The first parallel algorithm is called the "KJI Algorithm" and performs SAXPY operations on columns. Each SAXPY corresponds to a different column and hence can be parallelized over the processors.

Algorithm 9.1 Parallel KJI.

$$\textbf{for } k = 1, n - 1$$
```
CMIC$ PROCESS
```
$$A_{k+1:n,k} = (-1./A_{k,k}) * A_{k+1:n,k}$$
```
CMIC$ END PROCESS
CMIC$ DO GLOBAL
```
$$\quad \textbf{for } j = k + 1, n$$
$$\quad\quad A_{k+1:n,j} = A_{k+1:n,j} + A_{k+1:n,k} * A_{k,j}$$
$$\quad \textbf{end}$$
$$\textbf{end}$$

The second parallel algorithm is called the "Rank-Two Update Algorithm." It was parallelized for p processors by assigning roughly $1/p$ of the columns in the rank-two update operation to each processor.

Algorithm 9.2 Parallel Rank Two Update.

> for $k = 1, n, 2$
> CMIC$ PROCESS
> $\quad A_{k+1:n,k} = (-1./A_{k,k}) * A_{k+1:n,k}$
> $\quad A_{k+1:n,k+1} = A_{k+1:n,k+1} + A_{k,k+1} * A_{k+1:n,k}$
> $\quad A_{k+2:n,k+1} = (-1./A_{k+1,k+1}) * A_{k+2:n,k+1}$
> $\quad A_{k+1,k+2:n} = A_{k+1,k+2:n} + A_{k,k+2:n} * A_{k+1,k}$
> CMIC$ END PROCESS
> CMIC$ DO GLOBAL
> $\quad$ for $iproc = 1, p$
> $\quad\quad j1 = \lfloor(iproc - 1) * (n - k - 1)/p\rfloor + k + 2$
> $\quad\quad j2 = \lfloor iproc * (n - k - 1)/p\rfloor + k + 1$
> $\quad\quad A_{k+2:n,j1:j2} = A_{k+2:n,j1:j2} + A_{k,j1:j2} * A_{k+2:n,k}$
> $\quad\quad\quad\quad + A_{k+1,j1:j2} * A_{k+2:n,k+1}$
> $\quad$ end
> end

The third parallel algorithm is called the "Composite Algorithm" because it uses SMXPY operations on columns of L and SXMPY operations on rows of U. (SMXPY and SXMPY are "generalized" SAXPYs of Dongarra and Eisenstat [7] and are operations of the form $y \leftarrow y + Mx$ and $y^T \leftarrow y^T + x^T M$, respectively, where M is a matrix.) It was parallelized for two processors by assigning one processor to compute the U elements and the other processor to calculate the L elements. To adapt the Composite Algorithm for four processors, two are assigned for L calculations and two for U with vectors half as long. This approach can be generalized for a larger even number of processors but the shorter vector lengths would degrade its performance.

Note that the Cray microtasking directives PROCESS and ALSO PROCESS are used to demark processes which can be done in parallel. A partial pivoting strategy can be incorporated into these algorithms to obtain sound numerical procedures.

Table 9.4 gives timings in seconds for both the uniprocessor versions and microtasked versions of the Parallel KJI Algorithm, the Parallel Composite Algorithm, and the Parallel Rank-Two Update Algorithm. Representative results are given for a 700×700 dense random matrix using 1,2, and 4 processors on the four processor Cray X-MP4. Megaflop rates for each of the runs

Algorithm 9.3 Composite Algorithm for Two Processors.

```
for i = 2, n - 1
CMIC$ PROCESS
```
$$A_{i,i-1} = (-1./A_{i-1,i-1}) * A_{i,i-1}$$
$$\text{for } k = 1, i - 1$$
$$A_{i,i:n} = A_{i,i:n} + A_{i,k} * A_{k,i:n}$$
```
    end
CMIC$ ALSO PROCESS
```
$$A_{i+1:n,i-1} = (-1./A_{i-1,i-1}) * A_{i+1:n,i-1}$$
$$\text{for } k = 1, i - 1$$
$$A_{i+1:n,i} = A_{i+1:n,i} + A_{i+1:n,k} * A_{k,i}$$
```
    end
CMIC$ END PROCESS
end
CMIC$ PROCESS
```
$$A_{n,n-1} = (-1./A_{n-1,n-1}) * A_{n,n-1}$$
$$A_{n,n} = A_{n,n} + \langle A_{n,1:n-1}, A_{1:n-1,n} \rangle$$
```
CMIC$ END PROCESS
```

were computed and are given in parentheses below the timings. The speedup factors over one processor are given in parentheses under the megaflop rates.

From the numerical results in the table, the Parallel Composite Algorithm achieved good speedup ratios for two processors but less so for four processors (as expected). For the Parallel KJI Algorithm, the speedup factors using four processors was less effective than hoped for, due to excessive bank conflicts. The Parallel Rank-Two Update Algorithm behaved quite well based on speedup factors due to special coding in machine language. The maximum rate observed was 817 megaflops from a computer theoretically capable of a peak performance of 940 megaflops using four processors. Of the algorithms investigated in this experiment, the Parallel Composite Algorithm and the Parallel-Rank Two Update Algorithm were judged to be best for the architecture of the four processor Cray X-MP4 computer in both the uniprocessor and multiprocessor modes. These two algorithms had a common ingredient—kernels in which certain vectors are accumulation vectors. Thus, it was possible to use the vector registers as memory caches to hold intermediate results and thereby free the arithmetic pipes from interruptions due to the unavailability of vector operands. Dongarra, Gustavson, and Karp [8] have noted that the arithmetic for LU–factorization and other dense matrix

Table 9.4: Uniprocessor and Multiprocessor *LU*–Factorizations.

Method	n	Uniprocessor	Number of Processors		
			$p = 1$	$p = 2$	$p = 4$
KJI	700	1.56454	1.71196	0.88657	0.52578
		(146.0)	(133.4)	(257.6)	(434.4)
		(1.094)	(1.000)	(1.931)	(3.256)
Rank-Two	700	1.04488	1.04637	0.53092	0.27963
Update		(218.6)	(218.3)	(430.2)	(816.9)
		(1.001)	(1.000)	(1.971)	(3.742)
Composite	700	1.07445	1.07803	0.54826	0.29831
		(212.6)	(211.9)	(416.6)	(765.7)
		(1.003)	(1.000)	(1.966)	(3.614)

operations is free on the Cray computer; that is, designing an efficient algorithm on the Cray X-MP machine is a problem of memory management. It can also be concluded that there is not much room for improvement in the area of the parallel *LU*–factorization for full matrices on the Cray X-MP4 computer. The two best algorithms we investigated performed at essentially the top speed of this machine; any further improvement would have to come from efforts to more delicately balance the processors.

For additional information on these parallel algorithms, see the report by Oppe and Kincaid [24]. *LU*-factorization algorithms are the subject of a number of other papers (see, for example, Chen, Dongarra, and Hsiung [4] Dongarra, Gustavson, and Karp [8] and Dongarra and Hewitt [9]).

9.5 Conclusions

Parallel algorithms can be effectively implemented using microtasking on the four processor Cray X-MP4 supercomputer. In fact, the total execution time can be significantly reduced. Microtasking code can be expensive in terms of programming time; however, good speedup ratios are possible in many cases. (In some algorithms, speedups of 1.96 to 1.97 on two processors and 3.6 to 3.7 on four processors were obtained.) For each of the three investigations presented in this paper, the fastest execution time was obtained using four processors but the best speedup per number of processors was obtained with two processors.

Acknowledgments

We would like to thank Cray Research, Inc., for access to the Cray X-MP4 at Mendota Heights, MN. Also, the Cray X-MP2 of The University of Texas System was used and we wish to acknowledge its Center for High Performance Computing for assistance. This work was supported in part by the National Science Foundation under Grant DCR-8518722, by the Department of Energy under Grant DE-FG05-87ER25048, and Cray Research, Inc., under Grant LTR DTD with The University of Texas at Austin. Thomas C. Oppe's participation was also supported by Sandia National Laboratory through Contract No. 06-42982.

References

[1] Axelsson, O., *A Survey of Vectorizable Preconditioning Methods for Large Scale Finite Element Matrix Problems*, Report CNA-190, Center for Numerical Analysis, University of Texas at Austin, Texas, February, 1984.

[2] Calahan, D. A., "Task Granularity Studies on a Many-Processor Cray X-MP," *Parallel Computing*, 2, 109-118, 1985.

[3] Calahan, D. A., W. N. Joy, and D. A. Orbits, *Preliminary Report on Results of Matrix Benchmarks on Vector Processors*, University of Michigan Report SEL 94, May 1976.

[4] Chen, S. S., J. J. Dongarra, and C. C. Hsiung, "Multiprocessing Linear Algebra Algorithms on the Cray X-MP2: Experiences with Small Granularity," *Journal Parallel and Distributed Computing*, 1, 1, 22–31, 1984.

[5] Cray Computer Systems, *Multitasking User Guide*, Technical Note, Publication SN-0222, Revision B, March 1986.

[6] Dongarra, J. J., J. R. Bunch, C. B. Moler, and G. W. Stewart, LINPACK *User's Guide*, SIAM, Philadelphia, 1979.

[7] Dongarra, J. J., and S. C. Eisenstat, "Squeezing the Most Out of an Algorithm in CRAY Fortran," *ACM Trans. Math. Software*, 10, 3, 219–230, 1984.

[8] Dongarra, J. J., F. G. Gustavson, and A. Karp, "Implementing Linear Algebra Algorithms for Dense Matrices on a Vector Pipeline Machine," *SIAM Review*, 26, 1, 91–112, 1984.

[9] Dongarra, J. J., and T. Hewitt, "Implementing Dense Linear Algebra Algorithms Using Multitasking on the Cray X-MP4 (or Approaching the Gigaflop)," *SIAM Journal Sci. Stat. Comput.*, 7, 1, 347–350, 1986.

[10] Dongarra, J. J., and R. Hiromoto, "A Collection of Parallel Linear Equations Routines for the Denelcor HEP," *Parallel Computing*, 1, 2, 1984.

[11] Fong, K. W., and T. L. Jordan, *Some Linear Algebraic Algorithms and Their Performance on CRAY-1*, Los Alamos National Laboratory, Report LA-6774, June 1977.

[12] Hageman, L. A., and D. M. Young, *Applied Iterative Methods*, New York, Academic Press, 1981.

[13] Jordan, T. L., *A Performance Evaluation of Linear Algebra Software in Parallel Architectures*, Los Alamos National Laboratory, Report LA-8078-MS, October 1979.

[14] Jordan, T. L., "A Guide to Parallel Computation and Some Cray-1 Experiences," in *Parallel Computations*, G. Rodrigue (ed.), Academic Press, New York, 1982.

[15] Kincaid, D. R., and T. C. Oppe, "ITPACK on Supercomputers," *Numerical Methods*, V. Pereyra and A. Reinoza (eds.), *Lecture Notes in Mathematics* 1005, 151-161, Springer-Verlag, New York, NY, 1983. (Also Report CNA-178, Center for Numerical Analysis, University of Texas at Austin, Texas, September, 1982.)

[16] Kincaid, D. R., and T. C. Oppe, *A Parallel Algorithm for the General LU–Factorization*, Report CNA-208, Center for Numerical Analysis, University of Texas at Austin, Texas, April 1987.

[17] Kincaid, D. R., T. C. Oppe, J. R. Respess, and D. M. Young, *ITPACKV 2C User's Guide*, Report CNA-191, Center for Numerical Analysis, University of Texas at Austin, Texas, 1984.

[18] Kincaid, D. R., T. C. Oppe, and D. M. Young, "Adapting ITPACK Routines for Use on a Vector Computer," *Proceedings Symposium on Cyber 205 Applications*, The International Symposium on Vector Processing Applications, Colorado State University, Fort Collins, Colorado, August 12-13, 1982. (Also Report CNA-177, Center for Numerical Analysis, University of Texas at Austin, Texas, August, 1982.)

[19] Kincaid, D. R., J. R. Respess, D. M. Young, and R. G. Grimes. "ITPACK 2C: A Fortran Package for Solving Large Sparse Linear Systems by Adaptive Accelerated Iterative Methods." *ACM Trans. Math. Software*, 8, 302–322, 1982.

[20] Kincaid, D. R., and D. M. Young, *A Review of the ITPACK Project*, Report CNA-217, Center for Numerical Analysis, University of Texas at Austin, Texas, March 1988.

[21] Lawson, C. L., R. J. Hanson, D. R. Kincaid, and F. T. Krogh, "Basic Linear Algebra Subprograms for Fortran Usage," *ACM Trans. Math. Software*, 5, 3, 308–323, 1979.

[22] Moler, C. B., "Matrix Computations with Fortran and Paging." *Comm. ACM*, 15, 268–270, 1972.

[23] Oppe, T. C., and D. R. Kincaid. *Numerical Experiments with a Parallel Conjugate Gradient Method*, Report CNA-208, Center for Numerical Analysis, University of Texas at Austin, Texas, April 1987.

[24] Oppe, T. C., and D. R. Kincaid. *Parallel LU–Factorization Algorithms for Dense Matrices*, Report CNA-213, Center for Numerical Analysis, University of Texas at Austin, Texas, May 1987.

[25] Seager, M., "Parallelizing Conjugate Gradient for the CRAY X-MP," *Parallel Computing*, 3, 35–47, 1986.

[26] Stewart, G. W., *Introduction to Matrix Computation*, Academic Press, New York, 1973.

[27] van der Vorst, H. A., "The Performance of FORTRAN Implementations for Preconditioned Conjugate Gradients on Vector Computers," *Parallel Computing*, 3, 49–58, 1986.

Chapter 10

A SLAP for the Masses

Mark K. Seager*

10.1   Introduction

The Sparse Linear Algebra Package SLAP was conceived in 1985 in discussions with members of the Computing and Mathematics Research Division at Lawrence Livermore National Laboratory (Greenbaum [6]). It had three basic design criteria:

1. Ease of use for non-experts.

2. Ability to switch between methods with a minimum of effort.

3. Utility routines for various matrix manipulations and for viewing.

The package was first implemented, primarily by Anne Greenbaum, as a set of iterative and direct methods for use on Cray X/MP supercomputers. The original data structure (SLAP Triad format, described in Section 10.4) has the advantage that it is very simple to use, but has the significant drawback that none of the matrix vector operations vectorize. This version of SLAP also contained translation routines to various banded and sparse data structures so that one could use sparse direct methods as well. The user interface was simplified in that no workspace was required from the user,

---

*User Systems Division, Lawrence Livermore National Laboratory, Livermore, CA.

instead, workspace was obtained from the system using Cray Time Sharing System (CTSS) specific GETSPACE, RELSPACE memory allocation/deallocation routines and Cray FORTRAN compiler (CFT) pointer statements. In addition a matrix viewer routine was set up to utilize Livermore-specific interactive graphics output devices (TMDS). We got comments back from the users that they liked this type of user interface very much, but that it was slow (due to the non-vectorizability of the matrix vector operations) and was not portable (due to the memory allocation scheme and graphics output devices).

In 1987 a new set of design goals were established:

1. Portability and conformance to SLATEC (Fong *et al.* [5]) standards.

2. Ease of use for the non-expert.

3. Vectorization of matrix vector operations.

4. Ability to switch between iterative methods with a minimum effort.

5. Utility routines for matrix manipulation and viewing.

SLAP was rewritten to conform to these new design constraints by Anne Greenbaum and the author. First, the internal memory allocation (via pointers) was done away with and the necessary workspace obtained from the "user" via the subroutine calling sequence. This, of course, complicates the use of the package (prescribing the right amount of workspace for various routines is quite error prone, especially for non-experts), but until the FORTRAN standard is changed to handle some type of memory allocation this is the best one can do while retaining portability. Additional modifications to the package included consistent naming and internal documentation (FORTRAN comments) conventions based on the SLATEC standard. Finally, two new iterative methods were added for non-symmetric systems: generalized minimum residual (GMRES) (Saad and Schulz [10]) and bi-conjugate gradient squared (BCGS) (Sonneveld [12]) and the direct methods were dropped.

10.2   Data Structures

The core routines of SLAP are written in such a way that the data structure of the matrix and its associated preconditioner are only referenced in the "user supplied" routines MatVec and MSolve. Hence, the core routines are completely independent of the "user supplied" data structure. This allows the package great flexibility in that any method can be easily incorporated in a production code environment (i.e., it can conform to an existing highly structured data environment) if the user is willing to invest the effort of writing preconditioning and matrix vector multiply routines.

On the other hand, some code development environments do not so constrain programmers and they may choose to utilize directly a data structure supported by SLAP. In this situation the "high level" SLAP routines can be used with a choice of two matrix data structures. The simplest data structure is known as SLAP Triad format (also known as "coordinate format") and is illustrated in Figure 10.1. With this structure only non-zeros of the matrix are stored, in any order, in one **real** array A. If the matrix is symmetric only the lower triangle including the diagonal need be stored. Suppose **nelt** is the number of elements stored in A. Then two additional **integer** arrays, IA and JA, of length **nelt** are needed in the SLAP Triad format to hold the row and column indices of the matrix elements, respectively. This format is not very storage efficient, but it is trivial to set up. Unfortunately, the other drawback of this matrix storage mode is that the matrix vector multiply and incomplete factorization backsolves (preconditioning steps) do not vectorize. Therefore, if the user decides, for convenience, to use this matrix data structure the package transforms it automatically to the SLAP Column format.

$$
\begin{pmatrix}
a_{1,1} & a_{1,2} & 0 & 0 & a_{1,5} \\
a_{2,1} & a_{2,2} & 0 & 0 & 0 \\
0 & 0 & a_{3,3} & 0 & a_{3,5} \\
0 & 0 & 0 & a_{4,4} & 0 \\
a_{5,1} & 0 & a_{5,3} & 0 & a_{5,5}
\end{pmatrix} \Rightarrow
$$

Index	Element	Row	Column
1	$a_{5,1}$	5	1
2	$a_{1,2}$	1	2
3	$a_{1,1}$	1	1
4	$a_{3,3}$	3	3
5	$a_{1,5}$	1	5
6	$a_{5,3}$	5	3
7	$a_{5,5}$	5	5
8	$a_{2,2}$	2	2
9	$a_{3,5}$	3	5
10	$a_{4,4}$	4	4
11	$a_{2,1}$	2	1

(10.1)

Figure 10.1: SLAP Triad Matrix Storage Format.

The SLAP Column format is similar to the SLAP Triad format in that the non-zeros of the matrix are stored in a **real** array A and that the corresponding row indices are stored in an **integer** array IA. On the other hand, this data structure differs from the SLAP Triad format in that the matrix elements must be stored in a very specific order and the JA array has completely different interpretation. In the SLAP Column format all the non-zeros of the matrix elements must be stored by column starting with column 1 and

$$
\begin{pmatrix}
a_{1,1} & a_{1,2} & 0 & 0 & a_{1,5} \\
a_{2,1} & a_{2,2} & 0 & 0 & 0 \\
0 & 0 & a_{3,3} & 0 & a_{3,5} \\
0 & 0 & 0 & a_{4,4} & 0 \\
a_{5,1} & 0 & a_{5,3} & 0 & a_{5,5}
\end{pmatrix}
\Rightarrow
$$

Index	Element	Row	Column Pointer
1	$a_{1,1}$	1	1
2	$a_{2,1}$	2	4
3	$a_{5,1}$	5	6
4	$a_{2,2}$	2	8
5	$a_{1,2}$	1	9
6	$a_{3,3}$	3	12
7	$a_{5,3}$	5	
8	$a_{4,4}$	2	
9	$a_{5,5}$	5	
10	$a_{1,5}$	1	
11	$a_{3,5}$	3	

(10.2)

Figure 10.2: SLAP Column Matrix Storage Format.

ending with column n, where n is the number of unknowns in the problem. Secondly, the diagonal element must be the first entry stored for each column. The JA array is used in this context as offsets into the A and IA arrays for the the beginning of the compressed columns. In order to simplify coding, the last element JA(n+1) of JA, points to the beginning of column N+1, (which is "imaginary"). In other words, A(JA(k)) and IA(JA(k)) are the first stored matrix element and row index for the kth column and JA(n+1) points just past the last stored matrix element and index as seen in the example of Figure 10.2. An example of how to use this data structure for a matrix vector multiply routine is given in Figure 10.3. It is assumed that the matrix is symmetric and only the lower triangle and the diagonal are stored with isym = 1. Note that the inner loops, do 20 i = ibgn, iend and do 40 j = jbgn, jend, should vectorize on machines with hardware gather/scatter capabilities.

10.3 Preconditioners

Two preconditioners are supplied with the SLAP package. They both assume the SLAP Column format (although the "user level" routines will detect SLAP Triad format and automatically transform to the SLAP Column format). The first preconditioner supplied is the symmetric diagonal scaling (DS). Suppose we are solving the linear system

$$Ax = b. \tag{10.3}$$

Under diagonal scaling this system is transformed to

$$(\boldsymbol{D}^{-\frac{1}{2}}\boldsymbol{A}\boldsymbol{D}^{-\frac{1}{2}})(\boldsymbol{D}^{\frac{1}{2}}\boldsymbol{x}) = \boldsymbol{D}^{-\frac{1}{2}}\boldsymbol{b}, \tag{10.4}$$

where $\boldsymbol{D} = Diag(\boldsymbol{A})$ is the diagonal of $\boldsymbol{A}$. This corresponds to setting the preconditioning matrix $\boldsymbol{M} = \boldsymbol{D}$ and is the simplest of all preconditioners. It only requires that the diagonal be positive. The benefits of this preconditioner

```
      subroutine MatVec( n, x, y, nelt, ia, ja, a, isym )
      integer n, nelt, ia(nelt), ja(*), isym
      real x(n), y(n), a(nelt)
c
c         Compute y = A*x.
c
      do 10 i = 1, n
         y(i) = 0.0
 10   continue
      do 30 icol = 1, n
         ibgn = ja(icol)
         iend = ja(icol+1)-1
         do 20 i = ibgn, iend
            y(ia(i)) = y(ia(i)) + a(i)*x(icol)
 20      continue
 30   continue
      if( isym.eq.1 ) then
c
c         The matrix is symmetric.  Lower triangle is stored.
c         Multiply by the transpose, ignoring the diagonal.
c
         do 50 irow = 1, n
            jbgn = ja(irow)+1
            jend = ja(irow+1)-1
            if( jbgn.gt.jend ) GoTo 50
            do 40 j = jbgn, jend
               y(irow) = y(irow) + a(j)*x(ia(j))
 40         continue
 50      continue
      endif
      return
      end
```

Figure 10.3: Matrix Vector Multiply with SLAP Triad Format.

are: 1) it reduces the iteration count on most "real world" problems; and 2) it vectorizes very nicely.

The other preconditioner used in SLAP is modified incomplete LU factorization (LU) (modified incomplete Cholesky factorization (IC) if $\boldsymbol{A}$ is symmetric). Here, one chooses the preconditioning matrix to be the $\boldsymbol{LU}$ (or $\boldsymbol{LL}^T$ if $\boldsymbol{A}$ is symmetric) factorization of the original system without allowing any additional non-zeros to be created (no fill-in). It has been shown that if $\boldsymbol{A}$ is an M-matrix or an H-matrix (Manteuffel [9]) then the incomplete Cholesky factorization exists. In general the method can break down and SLAP then follows the strategy of Kershaw [8] by setting non-positive diagonal elements of the incomplete Cholesky factorization to unity. For most problems that are not too ill-conditioned this preconditioning reduces the iteration count substantially. On the other hand, the cost of computing the incomplete factorization is non-trivial and the vectorization of the backsolves (which must be done on each iteration) is necessarily done with short vectors and indirect addressing. Hence, modified incomplete factorization is not a clear winner over diagonal scaling for linear systems solved on vector supercomputers.

10.4   Iterative Methods

The SLAP package contains preconditioned iterative methods for both symmetric and non-symmetric linear systems. In Section 10.3 we discussed the preconditioners supplied with the package. As basic iterative methods for symmetric systems, one can choose from iterative refinement (IR), Jacobi (JAC), Gauss-Seidel (GS) and conjugate gradient (CG) iterative methods. The iterative refinement, Jacobi and Gauss-Seidel procedures are included mainly for comparison purposes and are not normally used for "real problems." The computational Conjugate Gradient algorithm used in this package is well known and can be found in Concus *et al.* [1]. See Table 10.1 for a tabulation of storage requirements and work estimates for these methods. In Table 10.1 and Table 10.2 the units of storage are: n (the number of unknowns), nl (the number of nonzeros stored in the lower triangle of the matrix, including the diagonal), nu (the number of nonzeros stored in the upper triangle of the matrix, including the diagonal), $nelt = nu + nl - n$ (the total number of nonzeros stored), MatVec (the number of operations required to perform a matrix vector multiply with the "user supplied" routine) and MSolve (the number of operations required to perform a preconditioning step with the "user supplied" routine). For the case where SLAP provides the MatVec and MSolve the work estimates are given in terms of n, nl, nu and $nelt$.

The largest number of methods in SLAP are provided for non-symmetric systems. If the A matrix is positive definite ($(Ax, x) > 0$ for all non-zero N-vectors x), then one can apply conjugate gradient on the normal equations (CGN) for

$$A^T A x = A^T b. \tag{10.5}$$

Rather than form $A^T A$ directly, which may not have a very sparse structure, we rewrite the conjugate gradient algorithm in terms of both A and A^T. This

Table 10.1: Storage and Work for Basic Iterative Methods.

Basic Methods				
Subroutine	Method	ISTORE	RSTORE	Work per Iteration
SIR	IR	0	3*n	2*n+MatVec+MSolve
SSJAC	JAC	10	4*n	2*n+2*nl
SSGS	GS	nl+n+11	nl+3*n	4*nl
SSILUR	ILU IR	nelt+3*n+11	nelt+3*n	4*n+6*nelt
SCG	CG	0	3*n	10*n+MatVec+MSolve
SSDCG	DS CG	10	4*n	10*n+2*nl
SSICCG	IC CG	nl+2*n+11	nl+5*n	8*n+4*nl

then requires two matrix vector multiplies per iteration (one of them is by the transpose of the matrix). Hence, for CGN (and most of the other methods for non-symmetric systems) the user must also supply a MTtVec routine (for the matrix transpose times a vector operation) as well as a MatVec routine. CGN has not been recommended in the past due to the fact that using the normal equations squares the condition number of the resulting iteration matrix. This causes this method to converge very slowly. Recently, this method has gained some popularity due to the fact that when a good preconditioner to $A^T A$ can be found this counteracts the effect of squaring the condition number, thereby making the method viable.

The Bi-conjugate gradient (BCG) method was proposed by Fletcher [4] for indefinite systems. It is similar to the conjugate gradient normal method in that it requires A^T as well as A, but it also requires the transpose of the preconditioner M^T. Hence, the user interface to bi-conjugate gradient requires four routines (MatVec, MTtVec, MSolve and MTSolve). The theoretical properties of bi-conjugate gradient are not very pleasing. In particular, if A is non-symmetric (the only case when one would use BCG) then bi-conjugate gradient is not guaranteed to reduce any quadratic functional (as in the case of conjugate gradient). In practice this rarely seems to be a problem and if one has a good preconditioner bi-conjugate gradient is an effective method.

For cases where the bi-conjugate gradient method converges, another algorithm developed by Sonneveld [12] converges twice as fast. This is the bi-conjugate gradient squared (BCGS) method. In addition to this property, bi-conjugate gradient squared also does not require A^T nor M^T and hence is much easier to use. It has been observed by the author and others that on problems where BCG is diverging BCGS will also diverge and twice as fast! In addition, on problems where bi-conjugate gradient seems to stagnate (i.e., not reduce the error for a large number of iterations) before converging, bi-conjugate gradient squared will more likely than not end up diverging. Moreover, the convergence and residual behavior for BCGS may be quite erratic. Hence, it is recommended that one use bi-conjugate gradient on the problem one wants to solve and if the method is converging switch to bi-conjugate gradient squared (with the same preconditioner). Due to the modular nature of SLAP, it is very easy to switch between methods in this manner.

Another extension of the conjugate gradient method to non-symmetric systems is orthomin (OMIN(k)) due to Vinsome [13]. For this algorithm one chooses the next search direction as a linear combination of the previous residual and k previous search directions so that the new search direction is A-orthogonal to the previous k search directions. If one minimizes the norm of the residual along this search direction, the orthomin algorithm is obtained (Jea and Young [7]). When k is chosen to be the iteration count (i.e. one A-orthogonalizes against all the previous search directions) then orthomin is guaranteed not to break down.

A Lanczos type extension of conjugate gradient for general non-symmetric systems can be found in the generalized minimum residual (GMRES) method of Saad and Schultz [10]. In this method, an orthonormal basis is generated via the Arnoldi process from the Krylov subspace

$$K(l) = span\{r_0, Ar_0, A^2 r_0, \ldots, A^{l-1} r_0\}, \tag{10.6}$$

where $r_0 = b - Ax_0$ is the initial residual. GMRES finds the approximate solution x_l in the affine subspace $x_0 + K(l)$ which has the minimal residual norm. This n-dimensional least squares problem can be reduced to a smaller l-dimensional least squares problem. GMRES is guaranteed to converge to the true solution in $l \leq n$ iterations for any non-singular matrix A. Usually however, a maximum value of l, denoted by k, is dictated by storage considerations to be very much smaller than n. If the stopping test is not met within k iterations, the iteration can be restarted by setting $x_0 = x_k$, applying the GMRES algorithm again, and so on. This algorithm is denoted by GMRES(k), and is guaranteed to converge as long as A is positive definite. A default value of $k = 10$ is used in the SLAP implementation of GMRES(k), but this value can be optionally set by the user. In building up the Krylov subspace $K(l)$ it is possible to estimate the error with $\frac{\|r_l\|}{\|b\|}$ without having to compute either the intermediate solution x_l or residual r_l. If one can utilize this "natural" GMRES(k) stopping test, then the algorithm is very efficient.

Detailed storage requirements and work estimates for the non-symmetric methods as implemented in SLAP are given in Table 10.2. As in Table 10.1, the storage requirements and work estimates are given in terms of n, k, MatVec and MSolve for the "core" methods and in terms of n, k, nl, nu

Table 10.2: Storage and Work for Basic Iterative Methods.

		Non-Symmetric Methods		
Subroutine	Method	ISTORE	RSTORE	Work per Iteration
SCGN	CGN	0	$3*n$	$13*n+2*MV+2*MS$
SSDCGN	DS CGN	10	$4*n$	$15*n+4*nelt$
SSLUCN	LU CGN	$nl+n+11$	$nl+3*n$	$17*n+12*n$
SBCG	BCG	0	$7*n$	$14*n+2*MV+2*MS$
SSDBCG	DS BCG	10	$8*n$	$16*n+4*nelt$
SSLUBC	LU BCG	$nl+nu+4*n+12$	$nl+nu+8*n$	$18*n+12*nelt$
SCGS	BCGS	0	$7*n$	$17*n+2*MV+2*MS$
SSDCGS	DS BCGS	10	$8*n$	$19*n+4*nelt$
SSLUCS	LU BCGS	$nl+nu+4*n+12$	$nl+nu+8*n$	$21*n+12*nelt$
SOMN	OMIN(k)	0	$n*(6+3*k)+k$	$(11+8*k)*n+MV+MS$
SSDOMN	DS OMIN(k)	10	$n*(7+3*k)+k$	$(12+8*k)*n+2*nelt$
SSLUOM	LU OMIN(k)	$nl+nu+4*n+12$	$nl+nu+n*(7+3*k)+k$	$(13+8*k)*n+6*nelt$
SGMRES	GMRES(k)	20	$n*(k+6)+k*(k+3)+1$	$3*k*n+MV+MS$
SSDGMR	DS GMRES(k)	30	$n*(k+7)+k*(k+3)+1$	$(1+3*k)*n+nelt$
SSLUGM	LU GMRES(k)	$nl+nu+4*n+32$	$nl+nu+n*(k+7)+k*(k+3)$	$(2+3*k)*n+2*nelt$

MV: MatVec
MS: MSolve

and *nelt* for the "high level" methods. The work description of GMRES(k) is not entirely accurate since the cost of restarting is not accounted for.

SLAP 2.0 iterative methods, matrix vector and preconditioner calculation routines follow a naming convention which, when understood, allows one to determine the iterative method, preconditioner and data structure(s) used in the routine. The subroutine naming convention takes the following form:

$$P[F][M]D,$$

where P stands for the precision (or data type) of the routine and is required in all names, the format code F denotes whether or not the routine requires the SLAP Triad or Column format (it requires the Column format if the second letter of the name is S otherwise it is matrix storage format independent), the optional M stands for the type of preconditioner used (this only appears in drivers for "core" routines) and D is some number of letters describing the method or purpose of the routine. In this version of SLAP both single and double precision data types are supported (although no complex data type routines have been written). Hence, all routines start with either the letter S or D. The brackets around S and M designate that these fields are optional.

The possibilities for the preconditioning, M, field are: D (diagonal scaling), IC (modified incomplete Cholesky factorization), ILU or LU (modified incomplete *LU* factorization). The description field, D, possibilities are: IR or R (iterative refinement), JAC (Jacobi), GS (Gauss Seidel), BCG or BC (biconjugate gradient), CG (conjugate gradient), CGS or CS (biconjugate gradient squared), GMRES, GMR or GM (generalized minimum residual), OMN or

OM (orthomin), DS (diagonal scaling preconditioner setup), D2S (diagonal scaling for normal equations preconditioner setup), 2LT (lower triangle preconditioning setup), ICS (incomplete Cholesky decomposition preconditioning setup), ILUS (incomplete LU decomposition setup), MV (matrix vector multiply), MTV (matrix transpose vector multiply), DI (SLAP solve for diagonal scaling), LI and LI2 (SLAP solve for lower triangle preconditioning), LLTI and LLTI2 (incomplete Cholesky SLAP solve), LUI and LUI2 (incomplete LU SLAP solve), LUTI and LUI4 (incomplete $(LU)^T$ SLAP solve), MMTI and MMI2 (SLAP solve for incomplete factorization preconditioning of the normal equations).

10.5   Utility Routines

The SLAP 2.0 package contains routines for manipulating data structures and for doing various matrix I/O. A short list of the routines and their purpose follows:

SBHIN   Single precision routine that reads in a sparse matrix in the Boeing/Harwell format.

QS2I2R   Sorts, using the quicksort algorithm, into ascending order an integer array carrying along one integer array and one real array. Can be used as the first step in transforming from the SLAP Triad format to the SLAP Column format.

SS2Y   SLAP Triad format to SLAP Column format converter.

SCPPLT   Printer plot of the SLAP Column format. For large matrices, only the first 132 rows and columns are displayed.

STOUT   Prints out a matrix, right hand side and solution (or any combination which includes the matrix) in the SLAP Triad format.

STIN   Reads in a matrix, right hand side and solution (or any combination which includes the matrix) in the SLAP Triad format.

10.6   Results

Four of the five problems (SHERMAN1, SHERMAN2, STEAM2 and JPWH991) used in these test came from the Harwell-Boeing sparse matrix collection (Duff *et al* [3]). SHERMAN1 is symmetric and arises from a three dimensional black oil reservoir simulator with shale barriers on a $10 \times 10 \times 10$

grid with one equation at each grid cell. SHERMAN2 is non-symmetric and also arises from three dimensional reservoir simulation. It is a problem that also involves simulation of steam injection into wells. The grid for this problem is 6 × 6 × 5 with 5 equations at each grid point. JPWH991 is non-symmetric and arises from circuit physics modeling. STEAM2 is non-symmetric and like SHERMAN1 arises from a three dimensional oil reservoir simulator. This problem models enhanced oil recovery by steam injection using a 5 × 5 × 5 grid with 4 variables at each grid cell. The only other problem with results listed below is the NASA1824 problem from H.D.Simon. This is a symmetric three dimensional structures problem. All these problems are considered difficult for both iterative and direct methods. For iterative methods, the preconditioner is of vital importance. For all the results given below, only the preconditioners supplied with the package (diagonal scaling and incomplete factorization) were used. No effort was made to tailor a preconditioner to any problem. This then can be considered the worse case for the iterative methods. Any production code developer would spend some amount of time adapting a preconditioner specialized to the problem being solved. In particular, for the three dimensional problems one would surely consider utilizing a preconditioner based on solving planes of two dimensional problems or some block factorization scheme.

Table 10.3: Vital Statistics for Sample Problems.

Problem Descriptions					
Problem	Size	NELT	SymNrm	% NZ	RCond
NASA1824	1,824	20,516	0.000	1.23	4.77E-7
SHERMAN1	1,000	3,750	0.000	0.38	2.17E-4
JPWH991	991	6,027	0.092	0.61	4.03E-3
SHERMAN2	1,080	23,094	0.995	1.98	1.68E-12
STEAM2	600	13,760	0.002	3.82	3.53E-7

All the above problems were solved with both direct methods (SGECO/SGESL and MA28) and iterative methods in the form of SLAP version 1.0 (the original Triad data structure, etc.) and version 2.0 (new column data structure, etc.). Table 10.3 gives the basic statistics for the test problems in terms of the linear system size (Size), number of non-zeros stored (NELT), the $SymNrm = \frac{\|A-A^T\|}{\|A\|}$ and percent nonzero $\%NZ = \frac{NELT}{Size^2}$ and the estimate of the condition number of the matrix $Rcond = \frac{1.0}{Condition(A)}$.

The direct method SGECO was chosen in order to obtain the condition number estimate as well as the factorization. The FORTRAN LINPACK implementation was used with hand coded level 1 BLAS. For the MA28 runs the pivoting flag $u = 1.0$ (which implies partial pivoting for numerical stability) was chosen so that the method was as robust as possible.

The problems were solved on the Cray Y/MP832 (SN1002) at the NASA Ames NAS facility and the Alliant FX/8 (medusa.llnl.gov) at the Lawrence Livermore National Laboratory utilizing only one processor. Both machines have hardware support for vector gather (or scatter) operations. In the tables below the Cray Y/MP (single precision, 64 bit) results are given on the left hand side and the Alliant FX/8 (double precision, 64 bit) on the right. The *Time* results are in CPU Seconds and the *ITER* column is the number of iterations taken to solve the problems via the iterative methods to a tolerance of 10^{-6}. The *Int* and *Real* columns give the amount of integer and real workspace (respectively) required by the various methods. These numbers include the matrix, preconditioners and workspace. The solution and right hand side are not included in these statistics (since all methods must store these in the same fashion). For these tests the number of past vectors stored for the Orthomin method was varied between eleven and one. For the GMRES method this parameter was varied between eleven and five.

Table 10.4: Time and Storage Results for NASA1824.

Results for NASA1824					
	Cray Y/MP832		Alliant FX/8		Storage	
Method	ITER	Time	ITER	Time	Int	Real
SGECO		25.57		548.79	1,824	3,328,800
MA28		27.28		1,069.03	265,237	193,239
SSDCG	1,387	9.36	1,383	238.23	22,351	29,636
sdcg	1,386	18.49	1,382	760.47		
SSICCG	-5		-5		65,208	70,668
iccg	264	7.53	260	257.11		
SSDBCG	1,407	20.75	1,399	454.14	22,351	35,108
dsbcg	1,410	39.79	1,402	1,520.42		
SSLUBC	280	8.01	269	181.36	70,681	76,140
ilubcg	281	17.14	269	531.01		
SSDCGS	1,463	19.83	1,232	1,240.31	22,351	35,108
SSLUCS	387	10.92	311	218.22	70,681	76,140

Table 10.4 gives the results for the symmetric system NASA1824. This problem is sufficiently ill conditioned that the conjugate gradient method applied to the normal equations is not effective (i.e., did not converge) with diagonal scaling and incomplete factorization as preconditioners. Also, the Orthomin and GMRES methods did not converge for this problem. It is interesting to note that the incomplete Cholesky factorization for NASA1824

broke down and had to be modified (Kershaw [8]. The modification then produced a preconditioned system $(M^{-1}A)$ that was not positive definite. This breakdown was noticed by SSICCG and the iteration was terminated (and hence the *ITER* column shows the error return code of -5), but the SLAP version 1.0 routine iccg did not fail and was in fact able to compute a solution (not guaranteed by the mathematical theory). It is clear from these results that the diagonal scaling preconditioner is not competitive for this problem. The best method (SSLUBC), discounting iccg whose convergence is a matter of luck, was 5.46 times faster than the direct solve and 11.01 times faster than the sparse method on the Cray Y/MP (similar improvements are observed on the Alliant FX/8). The superiority of SSLUBC solution technique is also displayed in the much smaller amount of storage required.

Table 10.5: Time and Storage for SHERMAN1.

	Results for SHERMAN1					
	Cray Y/MP832		Alliant FX/8		Storage	
Method	ITER	Time	ITER	Time	Int	Real
SGECO		4.26		310.42	1,000	1,001,000
MA28		0.59		14.41	26,124	18,752
SSILUR	514	1.88	472	44.30	11,513	10,501
ilur	514	1.51	472	50.15		
SSDCG	232	0.22	230	11.13	4,761	8,751
sdcg	232	0.25	230	14.77		
SSICCG	39	0.16	40	4.37	8,137	11,126
iccg	39	0.13	40	4.75		
SSLUCN	225	2.06	222	39.82	11,513	14,501
ilucgn	225	1.36	222	45.13		
SSDBCG	232	0.87	230	18.92	4,761	11,751
dsbcg	232	0.48	230	24.67		
SSLUBC	39	0.37	40	7.59	11,513	14,501
ilubcg	39	0.25	40	8.58		
SSDCGS	199	0.38	186	15.07	4,761	11,751
SSLUCS	29	0.22	30	5.84	11,513	14,501
SSDOMN(7)	506	0.56	506	37.63	4,761	31,758
dsomn(7)	506	0.73	506	47.62		
SSLUOM(3)	61	0.24	58	6.67	11,513	22,504
iluomn(3)	61	0.22	58	7.47		
SSDGMR(11)	799	0.84	769	56.16	4,781	21,906
SSLUGM(11)	55	0.23	48	6.25	11,533	24,656

The SHERMAN1 problem turned out to be negative definite and this fact was detected by the SLAP version 2.0 methods. The SLAP version 1.0 methods had various difficulties ranging from operand range error to simply not converging. Therefore, the problem was recast, yielding the results in

Table 10.5. For this problem the real winner was SSICCG. It was 27 times faster than SGECO and 3.41 times faster than MA28 on the Cray Y/MP because this problem is so sparse (about 3.8 nonzeros per column) implying very short vectors (average length of 3.8) for the column oriented approach of SLAP version 2.0. In this situation the preconditioning, matrix multiply and the incomplete factorization algorithms are slower than the Triad format scalar algorithms on the Cray Y/MP. Similar results were observed on the Cray X/MP416 (not presented here). Here is the first qualitative difference in the results obtained on the Cray computers and the Alliant FX/8. Even for these short vectors, the Alliant FX/8 vector gather/scatter hardware was fast enough to give the advantage to the column orientated vector algorithms of SLAP version 2.0. Undoubtedly, the cache system on the Alliant FX/8 played a major role.

The JPWH991 problem also turned out to be negative definite and the results summarized in Table 10.6 reflect the solution of the recast system. Overall the nonsymmetric iterative methods were very competitive for this problem (even the conjugate gradient applied to the normal equations SSDCGN and SSLUCN). The quickest (SLAP version 2.0) routine SSDCS on the Cray Y/MP was over 80 times faster than MA28 and over 92 times faster than SGECO. Again one can see the problems with short vectors on the Cray Y/MP, but not on the Alliant FX/8.

The SHERMAN2 problem was the hardest problem for the iterative methods, most of which could not solve the problem. The methods that did succeed worked quite well in comparison with the direct methods. The fastest routine was SSLUGM(11), beating MA28 and SGECO, on the Cray Y/MP, by a factor of 113 and 18, respectively. MA28 suffered from a great deal of fill-in and setting the pivoting parameter $u = 0.1$ had little effect.

The easiest problem to solve in this set is the STEAM2 problem (Table 10.8). The iterative methods are vastly superior on this problem and give an indication of their power when utilized with appropriate preconditioners. This problem has a block-banded matrix (multiple diagonals made up of 4×4 blocks). The diagonal block seems to be quite dominant. Hence, the incomplete LU preconditioner performance is spectacular and the diagonal scaling preconditioner is quite good. The best method is the SSDGMR(5) due to the fact that the incomplete factorization is not cheap to calculate. SSDGMR(5) is 525 times faster than MA28 and 250 times faster than SGECO on the Cray Y/MP. The poor showing of MA28 is due to a large amount of fill-in. Reducing the pivoting parameter $u = 0.1$ does not change the fill-in behavior.

Table 10.6: Time and Storage results for JPWH991.

Results for JPWH991					
	Cray Y/MP832		Alliant FX/8		Storage	
Method	ITER	Time	ITER	Time	Int	Real
SGECO		4.15		305.90	991	983,072
MA28		3.63		87.71	83,349	52,649
SSILUR	126	0.558	120	15.10	16,031	15,027
ilur	126	0.576	120	19.20		
SSDCGN	180	0.675	181	16.00	7,029	13,955
dscgn	180	0.591	181	28.10		
SSLUCN	31	0.350	30	7.80	16,031	18,991
ilucgn	31	0.309	30	9.74		
SSDBCG	40	0.151	37	3.39	7,029	13,955
dsbcg	40	0.133	37	5.92		
SSLUBC	16	0.194	14	4.01	16,031	18,991
ilubcg	16	0.173	14	5.18		
SSDCGS	23	0.045	27	2.59	7,029	13,955
SSLUCS	11	0.120	10	3.03	16,031	18,991
SSDOMN(7)	43	0.051	45	4.35	7,029	33,783
dsomn(7)	43	0.094	45	6.47		
SSLUOM(9)	14	0.099	14	2.88	16,031	50,715
iluomn(9)	14	0.090	14	3.78		
SSDGMR(9)	42	0.446	52	3.90	7,049	21,992
SSLUGM(11)	14	0.092	14	2.75	16,051	29,056

10.7 Future Parallel Extensions

Although the parallelization of SLAP 2.0 concentrated on vectorizing the inner loops of the matrix vector operations (*SIMD* parallelism), one could consider *MIMD* parallelism of the basic iterative methods as well. It has been found that the overhead associated with MicroTasking a tightly coupled iteration like preconditioned conjugate gradient are high, but tolerable for large linear systems [11]. This approach to parallelizing iterative methods is preferable from the standpoint of ease of coding and runtime overheads than MultiTasking on the Cray supercomputer. See [2] for a description of the terms and concepts used in this section.

As discussed in Section 10.4, the basic computational cost of the iterative methods are the matrix vector operations and the vector operations (SAXPY-like and SDOT-like reductions). In considering the parallelization of the SLAP iterative methods, these operations require attention. The following code segment implements a SAXPY operation when the strides are one (as is always the case in SLAP):

Table 10.7: Time and Storage results for SHERMAN2.

Results for SHERMAN2						
	Cray Y/MP832		Alliant FX/8		Storage	
Method	ITER	Time	ITER	Time	Int	Real
SGECO		5.39		392.32	1,080	1,166,400
MA28		34.42		1,223.92	289,938	177,514
SSLUBC	14	0.417	14	9.85	50,521	53,748
ilubcg	14	0.685	14	23.48		
SSLUCS	8	0.327	8	8.01	50,521	53,748
SSLUOM(11)	15	0.327	15	8.70	50,521	85,079
iluomn(11)	19	0.495	15	17.48		
SSLUGM(11)	10	0.305	11	7.35	50,541	64,704

```
      c              Origional SAXPY loop.
                DO 10 I = 1, N
                     SY(I) = SY(I) + SA*SX(I)
         10     CONTINUE
```

In parallelizing this operation one has one option: trade vector length for parallel processing. This is dangerous because if N is not large enough (i.e., we are not well out on the assymtotic part of the vectorization performance curve) then dividing it by four or eight (to get four or eight way parallelism) will degrade vector performance due to the shorter vector lengths. Since vectorization typically yields up to a factor of ten performance enhancement, coming down the vector speedup curve can be disastrous. With this caveat, there are several **MicroTasking** directives that can be used with Cray's **CFT** and **CFT77** compilers to chop up very long loops and parcel parts out to processors. For example: CMIC\$ DO GLOBAL, CMIC\$ DO GLOBAL LONG VECTOR, CMIC\$ DO GLOBAL BY *expression* and CMIC\$ DO GLOBAL FOR *expression* . The first method requires more user intervention than the others in that the loop must be broken up by hand and the CMIC\$ DO GLOBAL directive applied to the outer DO loop. With this technique the parallel SAXPY loop is written as:

```
      CMIC$ DO GLOBAL
                DO 20 IPROC = 1, NPROC
                     DO 10 I = IPROC, N, NPROC
                          SY(I) = SY(I) + SA*SX(I)
         10          CONTINUE
         20     CONTINUE
```

This has the advantage that it is very easy to code up and to read. The drawback is that if NPROC is a large power of two then memory bank conflicts can arise. To avoid this problem one may choose to break up the vectors into chunks with unit stride delimiting the inner loop by IBGN(IPROC) and IEND(IPROC):

Table 10.8: Time and Storage results for STEAM2.

Method	Cray Y/MP832 ITER	Cray Y/MP832 Time	Alliant FX/8 ITER	Alliant FX/8 Time	Storage Int	Storage Real
			Results for STEAM2			
SGECO		1.00		66.98	600	360,600
MA28		2.10		51.21	62,961	45,039
SSILUR	1	0.132	1	3.00	29,933	29,321
ilur	1	0.151	1	5.64		
SSDCGN	20	0.051	20	1.64	14,371	18,561
dscgn	20	0.155	20	6.97		
SSLUCN	1	0.140	1	3.14	29,933	31,721
ilucgn	1	0.169	1	6.35		
SSDBCG	8	0.020	8	0.75	14,371	18,561
dsbcg	8	0.063	8	3.03		
SSLUBC	1	0.138	1	3.23	29,933	31,721
ilubcg	1	0.164	1	6.28		
SSDCGS	5	0.008	5	0.54	14,371	18,561
SSLUCS	1	0.135	1	3.18	29,933	31,721
SSDOMN(7)	8	0.007	9	0.64	14,371	30,568
dsomn(7)	8	0.035	9	2.07		
SSLUOM(1)	1	0.132	1	2.93	29,933	32,922
iluomn(1)	1	0.151	1	5.55		
SSDGMR(5)	5	0.004	5	0.31	14,391	21,002
SSLUGM(5)	1	0.135	1	2.93	29,933	34,162

```
CMIC$ DO GLOBAL
      DO 20 IPROC = 1, NPROC
        DO 10 I = IBGN(IPROC), IEND(IPROC)
          SY(I) = SY(I) + SA*SX(I)
10      CONTINUE
20    CONTINUE
```

The next compiler directive, `CMIC$ DO GLOBAL LONG VECTOR`, tells the system to automatically chop up the `DO 10` loop into two loops with the inner loop having length of 64 (or less) and a `MicroTasked` outer loop. The `MicroTasked` outer loop is given `NREP = INT(N/64)+1` repetitions. The vector length of the inner loop is `MOD(N,64)` and hence it may be less than 64. This approach gives a large number of small granularity parallel chunks to run on multiple processors, thereby making load balancing less of a problem. On the other hand, the vector speeds attainable with fixed inner loop vector lengths of 64 can lead to reduced vector efficiency. For example, the original `SAXPY` loop compiled with CFT77 4.0 and run serially (no `MicroTasking`) under UNICOS 4.0 on the Cray Y-MP (SN1002 at NASA Ames with cycle time of 6.3 nano-seconds) yields an asymptotic rate of 220 MFLOPS (the

CAL coded library routine gets 260 MFLOPS) and the strategy of length 64 inner and additional outer gets (without any MicroTasking) 148 MFLOPS. That is, a decrease of 33% due to the fact that CFT77 4.0 can do "bottom storing" allowing some load/store of vector operands to be overlapped with computation.

The CMIC$ DO GLOBAL BY *expression* directive behaves the same way as CMIC$ DO GLOBAL LONG VECTOR except that the inner loop length is *expression* instead of 64. The CMIC$ DO GLOBAL FOR *expression* directive causes the loop to be divided into *expression* number of parallel parts. Thus, one can specify the inner loop vector length with BY *expression* or the number of outer loop repetitions with FOR *expression*, whichever is more convient for the user. These two directives allow one to increase the granularity of each parallel part (i.e., increase the inner loop vector length) at the expense of load balancing and tune the vector/parallel trade-off for a particular loop.

Due to the global summation variable in SDOT-like dot product vector reduction operations, there is no automatic way to break up the loop with Cray MicroTasking.

```
        c              Original SDOT loop.
              DOT = 0.0
              DO 10 I = 1, N
                  DOT = DOT + SX(I)*SY(I)
       10     CONTINUE
```

There is some help for this type of loop with the new generation of Cray parallelization tools known as AutoTasking. Cray's AutoTasking is based on the VAST II product of Pacific Sierra Research and has been adopted by many other vendors as well (Alliant and ETA among them). One problem with the way AutoTasking parallelizes this loop is the tacit assumption that the job will have all the processors on the machine allocated at runtime which therefore breaks up the loop into this many parallel parts. There is no AutoTasking directive to specify the amount of parallelism to utilize. This can lead to possible load balancing problems (e.g. on an eight processor Y-MP one would get eight parallel sections and yet the job might have only three or five processors allocated during the execution of the loop).

With these examples in mind, we can next consider the more difficult matrix vector operations. The SLAP Column format is going to be of little use, without modification, for parallelizing the matrix vector multiply (MatVec) operation. This is due to the following facts: 1) most sparse matrices, as we have seen in Section 10.6, have few non-zeros per column and hence we can not trade vector length for parallelism; 2) the vector algorithm used in SLAP for the MatVec operation requires that no row entry appear more than once within each vector update. This is guaranteed in SLAP because the vector update is done on a column-by-column basis. See the DO 20 loop in Figure 10.3. One possible solution to this problem is to give each processor working on the MatVec operation a **local** copy of the result vector and then have some scheme for combining the results at the end.

For our example here, let us consider a more specialized matrix data structure: a symmetric banded matrix arising from a nine-point difference stencil. In this case the appropriate MatVec algorithm does the matrix vector multiply by diagonals. Therefore, one can break up the computation of the result vector into parts as in the SAXPY example, above.

Diagonal scaling preconditioning can be parallelized in the same way. The incomplete factorization is more difficut and the reader is directed to [11] for more information.

With this background we now can appreciate the parallelization of the iteration loop of parallel preconditioned conjugate gradient. The parallel SAXPY-like loops and the necessary synchronization for the results from dot products are indicated in Algorithm 10.1. This comes in the form of critical regions (denoted by bold type) and by BARRIER SYNC's.

Algorithm 10.1 Iteration of Parallel PCG.

$$\text{FOR } k = 1(1)N \text{ DO}$$

$$r[ibp, ipe] = (r - \alpha t)[ipb, ipe]$$
$$err(iproc) = (r,r)[ipb, ipe]$$
$$\mathbf{err} = \sqrt{\sum_{i=1}^{\mathbf{nproc}} \mathbf{err(iproc)}}$$
$$\text{BARRIER SYNC 1}$$
$$\text{IF(} err/bnrm < \text{ TOL) DONE}$$
$$z[ipb, ipe] = (M^{-1}r)[ipb, ipe]$$
$$bkden = bknum$$
$$bknum(iproc) = (z,r)[ipb, ipe]$$
$$\mathbf{bknum} = \sum_{i=1}^{\mathbf{nproc}} \mathbf{bknum(iproc)}$$
$$\text{BARRIER SYNC 2}$$
$$\beta = bknum/bkden$$
$$p[ipb, ipe] = (z, \beta p)[ipb, ipe]$$
$$\text{BARRIER SYNC 3}$$
$$t[ipb, ipe] = (Ap)[ipb, ipe]$$
$$akden(iproc) = (p,t)[ipb, ipe]$$
$$\mathbf{bkden} = \sum_{i=1}^{\mathbf{nproc}} \mathbf{bkden(iproc)}$$
$$\text{BARRIER SYNC 4}$$
$$\alpha = bknum/akden$$
$$x[ipb, ipe] = (x + \alpha p)[ipb, ipe]$$

10.8   Conclusions

The SLAP version 2.0 is a significant improvement over the version 1.0 code in terms of robustness, flexibility, and speed. The latter improvement can be negated on the Cray X/MP and Y/MP class supercomputers when the number of non-zeros per column is quite small. The iterative methods presented here are quite competitive (even with very general preconditioners) with the direct methods on typical problems found in industry. Typical speed-ups of the iterative methods are between several and several hundred for the sample results. For the more difficult problems even more improvement would be obtained by utilizing specialized preconditioners.

Acknowledgments

Many people have made contributions to SLAP 2.0, but the author would especially like to thank Anne Greenbaum for the initail development of the package and the conversion work she did to make the package more portable. The original code for the GMRES routines was obtained from Peter Brown and Alan Hindmarsh. Peter Brown participated in the effort to mold the routines into the SLAP scheme. This work was supported in part by the Applied Mathematical Sciences subprogram of the Office of Energy Research, Department of Energy, by Lawrence Livermore National Laboratory under contract W-7405-ENG-48. Finally, NASA Ames NAS facility contributed non-trivial amounts of CPU time during the acceptance test of the Cray Y/MP.

References

[1] Concus, P., G. H. Golub, and D. P. O'Leary, "A Generalized Conjugate Gradient Method for the Numerical Solution of Elliptic Partial Differential Equations," in *Sparse Matrix Computations*, J. R. Bunch and D. J. Rose (eds.), Academic Press, New York, 309–332, 1976.

[2] Cray Research, *Multitasking User Guide*, Technical Report SN-0222, Cray Research, Inc., 1984.

[3] Duff, I. S., R. Grimes, and J. Lewis, "Sparse Matrix Test Problems," *ACM-TOMS*, 15, 1, 1–15, 1988.

[4] Fletcher, R., "Conjugate Gradient Methods for Indefinite Systems," in *Numerical Analysis*, G. A. Watson (ed.), Springer-Verlag, Lecture Notes in Mathematics 506, New York, 73–89, 1976.

[5] Fong, K. W., T. H. Jefferson, and T. Suyehiro, *SLATEC Common Mathematical Library Source File Format*, Technical Report UCRL-53313, Lawrence Livermore National Laboratory, 1982.

[6] Greenbaum, A., "Routines for Solving Large Sparse Linear Systems," *Tentacle News Letter*, Livermore Computing Center, Lawrence Livermore National Laboratory, Livermore, CA 94550, January 1986.

[7] Jea, K. C., and D. M. Young, "On the Simplification of Generalized Conjugate-Gradient Methods for Nonsymmetrizable Linear Systems," *Linear Algebra Appl.*, 52/53, 399–417, 1983.

[8] Kershaw, D. S., "The ICCG Method for the Iterative Solution of Systems of Linear Equations," *J. Comp. Phys.*, 26, 43–65, 1978.

[9] Manteuffel, T. A., "An Incomplete Factorization Technique for Positive Definite Linear Systems," *Math. Comp.*, 34, 473–497, 1980.

[10] Saad, Y., and M. H. Schultz, "GMRES: A Generalized Minimal Residual Algorithm for Solving Nonsymmetric Linear Systems," *SIAM J. Sci. Stat. Comput.*, 7, 3, 856–869, 1986.

[11] Seager, M. K., "Overhead Considerations for Parallelizing Conjugate Gradient," *Com. Appl. Num. Meth.*, 2, 2, 273–279, 1986.

[12] Sonneveld, P., *CGS, a Fast Lanczos-Type Solver for the Nonsymmetric Linear Systems*, Technical Report 84-16, Delft University of Technology, Department of Mathematics and Informatics, Julianalaan 132, 262B BL DELFT, 1984.

[13] Vinsome, P. K. W., "ORTHOMIN, An Iterative Method for Solving Sparse Sets of Simultaneous Linear Equations," Paper SPE 5739, *Symp. Numer. Simulation of Reservoir Performance of SPE of AIME*, Los Angeles, California, February 19-20, 1976.

Chapter 11

Parallel Direct Solution
of Sparse Linear Systems

*Esmond Ng**

11.1   Introduction

In this chapter, we consider direct methods for solving large sparse linear systems

$$Ax = b$$

on multiprocessor computer architecures, where A is an $n \times n$ nonsingular matrix. The basic approach is to decompose A into triangular factors

$$Q_r A Q_c = LU,$$

where Q_r and Q_c are some permutation matrices chosen to preserve sparsity and/or maintain numerical stability, and L and U are respectively lower and upper triangular matrices. With the triangular factorization, the solution to the linear system can be obtained by forward and back substitution sweeps solving $Lu = Q_r b$ and $Uv = u$, and then setting $x = Q_c v$.

In general the most expensive part of the solution process is the factorization of A. Thus much effort has been spent in designing efficient factorization algorithms for both sequential and parallel computers. The objective of this

*Mathematical Sciences Section, Oak Ridge National Laboratory, Oak Ridge, TN.

chapter is to provide an overview of some of the approaches and to discuss some of the issues in the design of effective parallel factorization algorithms. An outline of the contents is as follows. In Section 11.2, we briefly survey some of the effective sequential algorithms for solving sparse linear systems. Parallel algorithms are then described in Section 11.3. Tools for identifying and exploiting parallelism are introduced in Section 11.4, together with a discussion of related issues. Finally, some concluding remarks are provided in Section 11.5.

11.2 Sequential Algorithms for Sparse Linear Systems

There has been extensive research in the design of efficient sequential algorithms for the solution of large sparse linear systems. For example, George and Liu [17] contains an excellent discussion of most of the state-of-the-art methods for solving sparse symmetric positive definite systems and Duff *et al.* [5] has a detailed description of some methods for handling sparse nonsymmetric problems. The approach we consider in this chapter can be summarized as follows. There are basically four steps in the solution process:

1. *Ordering*:
   Compute permutations Q_r and Q_c so that L and U are sparse, where $LU = Q_r A Q_c$.

2. *Symbolic factorization*:
   Compute the structures of L and U. Set up a compact data structure for storing the nonzeros of L and U.

3. *Numerical factorization*:
   Input $Q_r A Q_c$ and compute L and U numerically. (Pivoting may be needed to ensure stability.) Store the nonzeros in the fixed data structure determined at step 2.

4. *Triangular solution*:
   Solve $Lu = Q_r b$ and $Uv = u$. Set $x = Q_c v$.

The approach stated above has been widely adopted for solving sparse symmetric positive definite systems [7,8,17], in which case Cholesky factorization is employed and is numerically stable by choosing the diagonal elements as pivots (Wilkinson [37]). Moreover, $Q_c = Q_r^T$ and $U = L^T$. Because of the fact that the factorization is stable without pivoting, the *structure* of L can therefore be determined solely from the structure of $Q_r A Q_r^T$, if we make the assumption that exact cancellation does not occur during numerical factorization. Once the structure of L is known, a compact data structure can then be set up to exploit the sparsity of L. There are efficient symbolic factorization algorithms for computing the structure of L and setting up the data structure (Sherman [35]). Then the numerical factorization and triangular solution can be performed using the fixed data structure.

The set of nonzeros introduced into L during numerical factorization is referred to as *fill*. The role of the permutation Q_r is to control the amount of fill in L. It is well known that the choice of Q_r can affect the sparsity of L drastically. This is illustrated by an example in Figure 11.1, in which $\hat{Q}_r A \hat{Q}_r^T$ is obtained from $Q_r A Q_r^T$ by reversing the ordering of rows and columns. Unfortunately, the general problem of finding the permutation that minimizes

$$
Q_r A Q_r^T = \begin{pmatrix} \times & \times & \times & \times & \times \\ \times & \times & & & \\ \times & & \times & & \\ \times & & & \times & \\ \times & & & & \times \end{pmatrix} \qquad L = \begin{pmatrix} \times & & & & \\ \times & \times & & & \\ \times & \times & \times & & \\ \times & \times & \times & \times & \\ \times & \times & \times & \times & \times \end{pmatrix}
$$

$$
\hat{Q}_r A \hat{Q}_r^T = \begin{pmatrix} \times & & & & \times \\ & \times & & & \times \\ & & \times & & \times \\ & & & \times & \times \\ \times & \times & \times & \times & \times \end{pmatrix} \qquad \hat{L} = \begin{pmatrix} \times & & & & \\ & \times & & & \\ & & \times & & \\ & & & \times & \\ \times & \times & \times & \times & \times \end{pmatrix}
$$

Figure 11.1: An example illustrating the effect of permutations on the sparsity of the Cholesky factors.

the number of nonzeros in L is NP-complete (Yannakakis [38]). Thus, we have to rely on heuristic strategies for finding permutations that reduce fill in of L. Some of the well-known strategies are the nested dissection algorithm [10,16] and the minimum degree algorithm [18]. Efficient implementations of these ordering algorithms and algorithms for the other three steps can be found, for example, in the SPARSPAK package (Chu *et al.* [2]).

For sparse nonsymmetric problems, it is well known that pivoting is necessary to ensure stability during numerical factorization (Wilkinson [37]). Since the choice of pivot at each step of the numerical factorization depends on both the structure and the numerical values of the active matrix, it is not clear how steps 1 and 2 can be performed prior to step 3. In fact, in almost all implementations of sparse triangular factorization with pivoting, steps 1, 2 and 3 are often combined together [3,25,27,36]. For example, at each elimination step in the routines MA28 of [3] (during which a row is eliminated), the pivot is chosen to preserve sparsity and to maintain stability. Then storage for the nonzeros is allocated immediately before the elimination of the row is performed numerically.

However, if we relax somewhat the condition that only nonzeros are stored, then it is possible to apply the previous four-step approach to nonsymmetric problems. Suppose $Q_r = Q_c = I$ for the moment. Consider computing a triangular factorization of A using Gaussian elimination with partial pivoting (i.e., row interchanges):

$$A = P_1 L_1 P_2 L_2 \cdots P_{n-1} L_{n-1} U,$$

where P_i corresponds to the row interchange that occurs at step i and L_i is a Gauss transformation at step i. Define $L = \sum_{i=1}^{n-1} L_i - (n-2)I$. George and Ng [21] have presented a symbolic factorization algorithm that will generate a lower triangular matrix $\bar{L}$ and an upper triangular matrix $\bar{U}$ from the *structure of A alone* so that the structures of $\bar{L}$ and $\bar{U}$ contain respectively those of L and U, *irrespective of the choice of P_i*, $1 \le i \le n - 1$. Thus, we can use the structures of $\bar{L}$ and $\bar{U}$ as bounds on the structures of L and U respectively in step 2 of the solution process. Using this approach, an effective static data structure for Gaussian elimination with partial pivoting can be set up [20]. Preserving the sparsity of $\bar{L}$ and $\bar{U}$ is important for the effectiveness of this scheme. It was demonstrated in [21] that the sparsity of $\bar{L}$ and $\bar{U}$ depends on the column ordering of A, and furthermore a good symmetric reordering of $A^T A$ appears to be a good column reordering of A.

We conclude the discussion in this section by presenting results of some numerical experiments. The objective is to illustrate the cost of performing each step in the solution process. There are two sets of test problems, all of which are finite element problems defined on L-shaped domains with triangular elements. The problems in the first set are symmetric positive definite and those in the second set are nonsymmetric (with symmetric structures). The experiments were performed on (one processor of) a Sequent Balance 8000 using single-precision floating-point arithmetic and execution times are in seconds. The symmetric positive definite problems were solved using the SPARSPAK package, and the nonsymmetric problems were solved using the approach described in [20]. The results are provided in Tables 11.1 and 11.2.

Table 11.1: Execution time statistics (in seconds) for symmetric positive definite problems. (n is the order of the matrix.)

n	3025	3466	3937	4438	4969
ordering	8.550	9.917	11.533	13.183	15.100
symbolic factorization	1.917	2.150	2.450	2.783	3.117
numerical factorization	59.533	73.317	89.167	105.817	126.567
triangular solution	4.867	5.733	6.550	7.617	8.667

Table 11.2: Execution time statistics (in seconds) for nonsymmetric problems. (n is the order of the matrix.)

n	3025	3466	3937	4438	4969
ordering	48.800	57.266	69.883	78.483	89.367
symbolic factorization	4.717	5.450	6.117	7.017	7.933
numerical factorization	502.050	646.283	776.300	915.917	1165.133
triangular solution	12.300	14.650	16.950	19.350	22.733

11.3 Parallel Sparse Matrix Factorization Algorithms

It is clear from the numerical results in the previous section that the ordering, symbolic factorization and triangular solution phases are relatively inexpensive; the numerical factorization phase is usually the most expensive part of the solution process. Thus, when multiprocessor systems became available, much effort was spent on parallelizing numerical factorization. In this section, we discuss the potential sources of parallelism in sparse numerical factorization.

We begin with a description of a sequential numerical factorization algorithm, in which columns are eliminated, and we will ignore sparsity for the moment:

```
for k = 1 to n
    perform row and/or column interchanges, if necessary
    compute multipliers at step k
    for j = k + 1 to n
        modify row/column j by row/column k
```

Whether we use a row-oriented algorithm or a column-oriented algorithm will depend on the choice of data structure for the matrix A. For example, if the elements of A are stored by columns, then it may be beneficial to use the column-oriented algorithm to facilitate access of the matrix elements. At any rate, we see from the algorithm above that the computation at each major step can be broken up into subtasks: $cdiv(k)$ and $mod(j, k)$. The subtask $cdiv(k)$ refers to the computation of the multipliers at step k, which includes the row and/or column interchanges that may be necessary to ensure numerical stability, and the subtask $mod(j, k)$ is the modification of row/column j by row/column k. Thus, we can express the computation in a compact way:

> **for** $k = 1$ **to** n
> $cdiv(k)$
>   **for** $j = k + 1$ **to** n
> $mod(j, k)$

It is important to note that for a given k, each $mod(j, k)$ subtask uses data from row/column k to update row/column j. Hence, the $mod(j, k)$ subtasks are *independent* subtasks for a fixed k. Suppose there are several processors available in a multiprocessor system. As long as $cdiv(k)$ has been performed and row/column k is available to each processor, the independent subtasks $mod(j, k)$ may therefore be performed concurrently. Note that the operations within a *mod* subtask are also independent. Such independence may be exploited, for example, if the processors have vector processing capability or if there are enough processors in the multiprocessor system. However, we will not consider such fine-grain parallel algorithms in this chapter; we are interested in medium-grain parallel algorithms. Furthermore, note that the $cdiv(k)$ subtask cannot begin until $mod(k, i)$ has been performed for all $i < k$. Thus, if the matrix is dense, the $cdiv$ subtasks will be executed sequentially even though some of the *mod* subtasks can be carried out in parallel.

Sparsity of the matrix can enhance the amount of parallelism available. As an illustration, let A have the form

$$A = \begin{pmatrix} \times & & & & & \times & & \times & \\ & \times & & & & & \times & & \\ & & \times & & & & & \times & \\ & \times & & \times & & & & & \\ & & \times & & \times & & \times & & \\ & \times & & & & \times & & & \times \\ & & \times & & \times & & \times & & \\ & & & \times & & \times & & \times & \\ & \times & & & & \times & & \times & \\ & & & & \times & \times & & \times & \end{pmatrix}$$

For definiteness, suppose we are using a row-oriented factorization algorithm. Consider the first three steps of the factorization, and for simplicity ignore the necessity of pivoting. Because of the nonzero pattern, the first three columns are independent. This implies that $cdiv(i)$, $1 \leq i \leq 3$, can be carried out simultaneously, provided that there are enough processors available. This small example illustrates the fact that because of sparsity in the matrix, not only can some of the *mod* subtasks be executed in parallel, but some of *cdiv* subtasks may become independent during factorization, and they can be performed concurrently.

Parallel sparse numerical factorization algorithms have been developed for various classes of multiprocessor systems [4,6,12,14,15,22,23]. Of course, the crucial issue is how the independence among the *cdiv* and *mod* subtasks can be identified and how such independent computations can be scheduled in such a way that the computational load is balanced and the amount of synchronization or communication is kept low. We will discuss these issues in the next section.

We have concentrated our discussion in this section on the potential for parallelism in sparse numerical factorization. We conclude the section by making a few remarks about parallel algorithms for the other phases in the solution process. Although the numerical factorization phase is usually the most expensive phase in the solution process on sequential machines, if we are able to reduce the factorization time by employing multiple processors on a multiprocessor system, the cost of doing the remaining three phases (sequentially) may become significant. Thus, research on designing efficient parallel algorithms for the ordering, symbolic factorization and triangular solution phases has been initiated. Some of the work has been reported in [1,12,13,14,24,26,39,40]. However, since the amount of computing in ordering, symbolic factorization or triangular solution is often relatively small, and the sequential algorithms for each of these three phases are extremely efficient, it is in general difficult to devise parallel algorithms with good efficiencies for these three phases. On the other hand, there are situations in which parallel algorithms for ordering, symbolic factorization and triangular solution are desirable even though efficiencies may be poor. For example, if the numerical factorization is performed on a local-memory multiprocessor, the columns of L will be distributed among the processors. Instead of sending the columns from the processors to a single processor and performing the triangular solution sequentially, it may be desirable to have all the processors collaborate in the computation.

11.4 Identifying Parallelism and Scheduling Independent Subtasks

In the previous section, we have described the potential parallelism available in sparse numerical factorization. However, in order to make use of multiple processors in the numerical factorization, it is important to have efficient

$$
A = \begin{pmatrix}
\times & & & \times & & \times & & & & \times & \\
& \times & \times & & & & \times & & \times & & \\
& \times & \times & & & & \times & & & \times & \\
& & & \times & & & \times & & & & \times \\
\times & & & & \times & & & & \times & & \\
& & & \times & \times & & \times & & & & \times \\
& \times & \times & \times & & \times & \times & & \times & & \\
& & & & \times & & & \times & & & \times \\
\times & & & & \times & & & & \times & & \times \\
& \times & & & & \times & & & & \times & \\
& & & \times & & & \times & & \times & & \times
\end{pmatrix}
$$

$$
L = \begin{pmatrix}
\times & & & & & & & & & & \\
& \times & & & & & & & & & \\
& & \times & & & & & & & & \\
& & \times & \times & & & & & & & \\
\times & & & & \times & & & & & & \\
& & \times & \times & & \times & & & & & \\
& & \times & & + & & \times & & & & \\
& & & \times & & \times & + & + & \times & & \\
\times & & & & + & & \times & + & + & \times & \\
& & & & & & \times & \times & + & \times
\end{pmatrix}
$$

Figure 11.2: A sparse symmetric positive definite matrix and its Cholesky factor. ($\times$ is a nonzero in the original matrix and $+$ is a fill element.)

tools for identifying independent subtasks in the computation and scheduling these subtasks among the processors. For simplicity, let us consider the case in which the matrix A is symmetric and positive definite. Denote by L the Cholesky factor of A. Thus, $A = LL^T$ and L is lower triangular. If column k of L has more than one nonzero, then let $f(k)$ be the row index of the first off-diagonal nonzero; otherwise, let $f(k) = 0$. Then $\{f(k)\}$ forms a *tree* structure, which is often referred to as the *elimination tree* of A. More precisely, if $f(k) \neq 0$, then node $f(k)$ is the *parent* of node k in the elimination tree and k is one of several children of $f(k)$. An example is given in Figures 11.2 and 11.3. In general, there may be several disjoint elimination trees associated with A. However, there is exactly one elimination tree if A is irreducible. In our discussion below, we will assume that the matrix A is irreducible. Each elimination tree has a distinct node $\hat{k}$, called the *root*, such that $f(\hat{k}) = 0$. There is a unique path between the root of the elimination tree and any node in the same tree. Suppose nodes i and j are in the same elimination tree, with $i < j$. If node j is on the path between the root and node i, then node j is an *ancestor* of node i and node i is a *descendant* of node j. A *subtree rooted at node* i is the set of all descendants of node i in the elimination tree.

The notion of elimination trees has been used extensively in research on sparse direct methods, and an excellent survey can be found in [31]. For

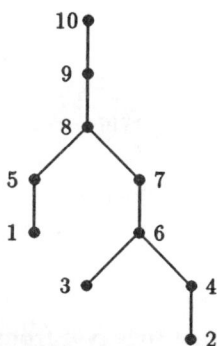

Figure 11.3: The elimination tree associated with the matrix in Figure 11.2.

example, elimination trees are employed in [4,6,11,14,15,20,22,28,30,32,39,40].
Other references related to elimination trees are in [31]. It should be noted
that the *structure* of the elimination tree depends solely on the structure of A.
Moreover, the elimination tree can be computed directly from the structure
of A.

The elimination tree of A provides much information about the dependency among the *cdiv* and *mod* subtasks in sparse numerical factorization.
It serves as a tool for identifying and exploiting parallelism in sparse matrix factorization. Consider Cholesky factorization and suppose we are using a column-oriented algorithm. We note that in order to perform $cdiv(k)$,
$mod(k,i)$ has to be performed first, for all $i < k$ such that $L_{ki} \neq 0$. It is
easy to show that node i must be a descendant of node k in the elimination
tree; the proof follows from the way the Cholesky factor L is computed. A
corollary of this result is that column k of L depends explicitly on column k
of A *and* a subset of the columns of L that are associated with the subtree
rooted at node k. In other words, $cdiv(k)$ cannot be executed unless

1. $cdiv(i)$ has been performed, for all nodes i in the subtree rooted at node
   k, and

2. $mod(k,i)$ has been applied, for appropriate nodes i in the subtree rooted
   at node k.

In general, column k of L depends either explicitly or implicitly on the
columns of L that are associated with the subtree rooted at node k. Hence,
for $k_1 \neq k_2$, if nodes k_1 and k_2 are in two *disjoint* subtrees, then columns k_1
and k_2 are independent, since the two sets of columns on which columns k_1
and k_2 depend are disjoint. For the example in Figures 11.2 and 11.3, columns
1, 2 and 3 are therefore independent. Also, columns 5 and 7 are independent,

but column 7 depends either explicitly or implicitly on columns 2, 3, 4 and 6. In summary, dependencies among subtasks in sparse numerical factorization can be identified by analyzing the structure of the elimination tree associated with the matrix A.

Since the elimination tree provides information on the dependency among the subtasks in sparse numerical factorization, the tree can be used to schedule the independent computations on a multiprocessor system. One strategy is to schedule the columns by pruning the elimination tree. The idea is to schedule the columns so that the *cdiv*'s can be performed as soon as possible. For example, for the matrix in Figure 11.2, we will first schedule columns 1, 3 and 2. This amounts to removing the leaves from the elimination tree. Then based on the pruned tree, we schedule columns 5 and 6, and again this corresponds to removing the leaves from the pruned tree. The process is repeated until all the columns are scheduled. Note that the height of the elimination tree is a lower bound on the number of serial *cdiv*'s that have to be performed in a parallel sparse numerical factorization algorithm.

The elimination tree is also useful in reducing the amount of communication or synchronization required in parallel sparse numerical factorization. Observe that if the columns of a subtree are assigned to a subset of processors, then the communication or synchronization involved when these columns are computed will be limited to the processors in this subset, although communication or synchronization may be required in order to make these columns available to those associated with the ancestors of the subtree. Thus, by assigning disjoint subtrees to different disjoint sets of processors, communication or synchronization requirements are reduced. George, Liu and Ng have used this observation to assign columns to processors to reduce the cost of communication in the numerical Cholesky factorization on multiprocessor systems with the hypercube topology [19].

The discussions above demonstrate that effective parallel sparse numerical factorization relies on the structure of the elimination tree. For example, a balanced elimination tree with many branches appears to be desirable. So what is a *good* elimination tree for parallel sparse numerical factorization? Note that the elimination tree is defined in terms of the structure of the Cholesky factor of A. Since we know that the structure of the Cholesky factor depends on the reordering of columns and rows of A, the structure of the elimination tree depends on the row and column reordering. An example illustrating the effect of reordering on the structure of the elimination tree is provided in Figures 11.4 and 11.5, in which $\bar{A}$ is obtained from the matrix A in Figure 11.2 by reversing the ordering of rows and columns. Thus, we can rephrase the question as follows. Given the structure of a matrix, what is a good reordering for parallel sparse numerical factorization?

$$
\bar{A} = \begin{pmatrix}
\times & & \times & \times & & & & & & \\
& \times & & & \times & & & & & \times \\
\times & & \times & & & \times & & \times & & \\
\times & & & \times & & \times & & & & \times \\
& \times & \times & & & \times & & & & \times \\
& & & \times & \times & & \times & & \times & \\
& & \times & \times & & & \times & & \times & \\
& & \times & & & \times & & \times & & \times \\
& & & \times & & & \times & & \times & \times \\
& \times & & & \times & & & \times & & \times
\end{pmatrix}
$$

$$
\bar{L} = \begin{pmatrix}
\times & & & & & & & & & \\
& \times & & & & & & & & \\
\times & & \times & & & & & & & \\
\times & & + & \times & & & & & & \\
& \times & & & \times & & & & & \\
& & \times & + & & \times & & & & \\
& & & & \times & & \times & & & \\
& & \times & + & \times & + & + & \times & & \\
& & & \times & & & + & \times & + & \times \\
& & \times & & + & \times & + & + & + & \times
\end{pmatrix}
$$

Figure 11.4: A symmetric matrix $\bar{A}$ and its Cholesky factor. ($\bar{A}$ is obtained by reversing the ordering of the rows and columns of the matrix A in Figure 11.2.)

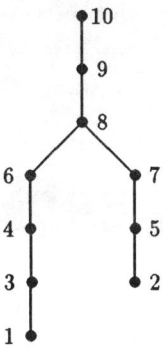

Figure 11.5: The elimination tree associated with the matrix $\bar{A}$ in Figure 11.4.

Since the height of the elimination tree is a lower bound on the number of serial *cdiv*'s that have to be performed, it is desirable to find a reordering for A so that the elimination tree is as short as possible. However, Pothen has shown that the problem of finding such a reordering is NP-complete [34].

Recall that the sparsity of L also depends on the choice of the reordering. Hence, it is desirable not only to find a reordering so that the associated elimination tree is short, but also the fill is small. This suggests the following heuristic. We first determine a reordering P_r that attempts to minimize fill in L, where $LL^T = P_r A P_r^T$. Thus, there is an elimination tree associated with $P_r A P_r^T$. Then the tree is restructured in such a way that the new tree has a different shape and hopefully has a smaller height, but the fill and operation count are preserved. This approach is proposed by Liu [30,32]. It amounts to finding another reordering $\bar{P}_r$ for $P_r A P_r^T$ with the constraint that both fill and operation count are preserved. When fill and operation count are preserved, the reordering $\bar{P}_r$ is said to be *equivalent* to P_r.

If A is a finite element matrix arising from a two-dimensional problem, and P_r is a nested dissection reordering, then our experience is that the resulting elimination tree is often short and balanced, and more importantly, P_r often adequately reduces fill. Thus, nested dissection reorderings are often good reorderings for parallel sparse numerical factorization, at least for certain classes of problems. For general sparse symmetric positive definite problems, a minimum degree reordering is in general a much better reordering in terms of fill-reduction. Unfortunately, the resulting elimination tree is often tall and unbalanced, and hence, may not be suitable for parallel factorization. The example in Figure 11.6 is an elimination tree associated with a minimum degree reordering on a 7 × 7 grid, with a nine-point operator. When we apply Liu's heuristic to the elimination tree in Figure 11.6, we obtain the elimination tree in Figure 11.7, which is again not balanced, although the height of the new tree is somewhat smaller than that of the old tree. In fact, it is our experience that such a phenomenon is typical for minimum degree reorderings.

There is a different way of finding an elimination tree with a short height. Suppose we first find a reordering P_r that attempts to minimize fill in L, where $LL^T = P_r A P_r^T$. There may be equivalent reorderings which preserve fill and operation count. Each equivalent reordering will result in a Cholesky factor with different structure from that of L, and consequently a different elimination tree. Thus, a possibility is to choose from the set of equivalent reorderings the one that has an elimination tree of the shortest height. The scheme for finding the equivalent reordering was first described by Jess and Kees [28], but it was Liu who proved that the resulting elimination tree indeed has the shortest height [30]. Implementations of this scheme were described in [29] and [33]. Unfortunately, we do not have much numerical experience with this approach, since the codes are not available to us. However, even though the new reordering gives an elimination tree that has the smallest height, there is no guarantee that the new tree is balanced or has many branches.

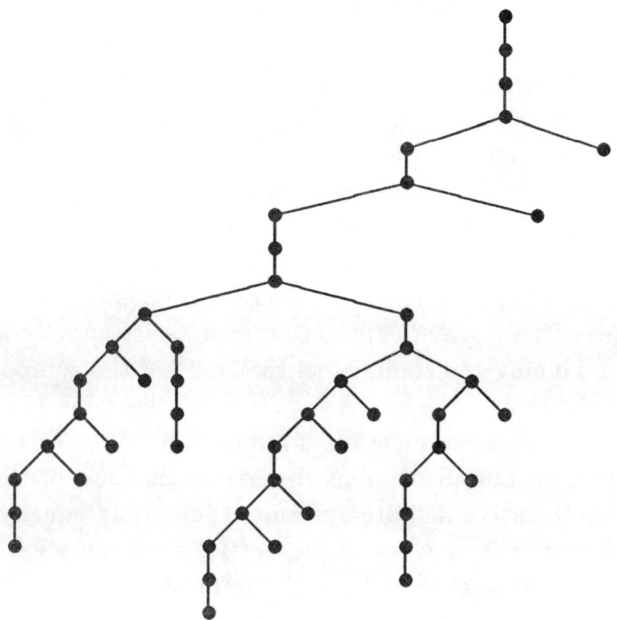

Figure 11.6: The elimination tree associated with a minimum degree reordering on a 7 × 7 grid.

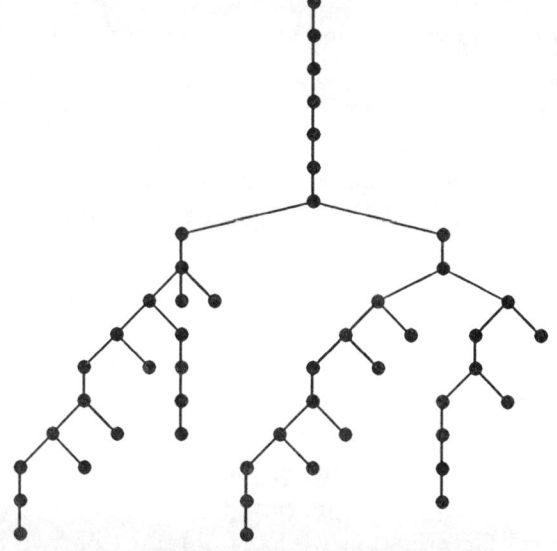

Figure 11.7: The elimination tree obtained by applying Liu's heuristic to the elimination tree in Figure 11.6.

Investigation into the effectiveness of this technique in terms of parallel sparse numerical factorization is underway.

The height of an elimination tree is not the only criterion for effective parallel sparse numerical factorization. The shape of the elimination tree is also important. It is our experience that elimination trees which are balanced and have many branches, such as those corresponding to nested dissection orderings, appear to be desirable. However, for most reorderings, their elimination trees are not balanced. Thus, for unbalanced elimination trees, another issue is how to schedule the columns so that the computational work is balanced and the synchronization or communication requirements are reduced. Based on the structure of a *weighted* elimination tree, Geist and Ng have proposed a heuristic for assigning the columns so that part of the computation can be distributed evenly among the processors, and at the same time the amount of synchronization or communication is reduced [9].

Finally, although our discussions in this section are on the solution of sparse symmetric positive definite systems, they apply equally well to nonsymmetric problems. When A is nonsymmetric, the elimination tree is defined in terms of the structure of the Cholesky factor of $A^T A$ [20,22,23].

11.5 Numerical Experiments and Concluding Remarks

In this chapter, we have provided an overview of the current state of affairs in the direct solution of sparse linear systems on multiprocessor systems. Clearly one of the open problems is how to characterize and determine an appropriate reordering for parallel sparse numerical factorization. Note that there are several constraints to be satisfied. It is important to find a reordering so that fill is reduced, and it is desirable to choose the reordering so that the height of the resulting elimination tree is minimized. Moreover, it is also important to have an elimination tree that contains many branches and is balanced, so that there is a high degree of parallelism. Of course, to make the problem even harder, it is desirable to be able to compute the reordering itself in parallel. There are other problems to consider as well, such as the design of efficient parallel algorithms for performing symbolic factorization and triangular solution. These problems are under investigation.

We conclude this chapter by providing some numerical results for parallel sparse numerical factorization we have obtained on two different parallel computers. The test problems are the finite element problems we have used in Section 11.2. Tables 11.3 and 11.4 contain respectively the performance statistics for sparse symmetric positive definite problems and sparse non-

Table 11.3: Performance results on a Sequent Balance 8000 for sparse symmetric positive definite systems. For each problem/processor pair, the three entries are respectively the execution time in seconds, the speed-up ratio and the efficiency.

n	sequential	$p = 2$	$p = 4$	$p = 6$
3025	59.533	37.700	19.967	14.333
		1.58	2.98	4.15
		78.96%	74.54%	69.23%
3466	73.317	45.967	24.200	17.183
		1.59	3.03	4.27
		79.75%	75.74%	71.11%
3937	89.167	55.783	29.383	20.700
		1.60	3.03	4.31
		79.92%	75.87%	71.79%
4438	105.817	66.233	34.633	24.467
		1.60	3.06	4.32
		79.88%	76.38%	72.08%
4969	126.567	78.533	41.133	29.033
		1.61	3.08	4.36
		80.58%	76.93	72.66%

Table 11.4: Performance results on a Sequent Balance 8000 for sparse nonsymmetric systems. For each problem/processor pair, the three entries are respectively the execution time in seconds, the speed-up ratio and the efficiency.

n	sequential	$p = 2$	$p = 4$	$p = 6$
3025	502.050	290.967	151.417	104.800
		1.73	3.32	4.79
		86.27%	82.89%	79.84%
3466	646.283	376.300	194.933	133.833
		1.72	3.32	4.83
		85.87%	82.89%	80.48%
3937	776.300	449.050	231.700	160.983
		1.73	3.35	4.82
		86.44%	83.76%	80.37%
4438	915.917	529.800	273.333	186.867
		1.73	3.35	4.90
		86.44%	83.77%	81.69%
4969	1165.133	666.150	343.517	235.750
		1.75	3.39	4.94
		87.45%	84.79%	82.37%

Table 11.5: Performance results on an Intel/iPSC-2 for sparse symmetric positive definite systems. For each problem/processor pair, the three entries are respectively the execution time in seconds, the speed-up ratio and the efficiency.

n	sequential	$p = 8$	$p = 16$	$p = 32$
2233	17.803	6.500	4.695	3.513
		2.74	3.79	5.07
		34.24%	23.70%	15.84%
2614	22.512	8.480	6.198	4.400
		2.65	3.63	5.12
		33.18%	22.70%	15.99%
3025	28.183	10.532	7.456	5.292
		2.68	3.78	5.33
		33.45%	23.62%	16.64%
3466	34.607	13.141	8.980	6.313
		2.63	3.85	5.48
		32.92%	24.09%	17.13%
3937	42.368	15.175	10.670	7.565
		2.79	3.97	5.60
		34.90%	24.82%	17.50%
4438	50.379	18.224	12.551	8.437
		2.76	4.01	5.97
		34.56%	25.09%	18.66%
4969	59.932	21.868	14.498	9.872
		2.74	4.13	6.07
		34.26%	25.84%	18.97%

symmetric problems on a Sequent Balance 8000, which is a multiprocessor system with shared-memory. Table 11.5 contains the performance statistics for sparse symmetric positive definite problems on an Intel/iPSC-2, which is a distributed-memory parallel machine. For each sparse symmetric positive definite problem, a nested dissection reordering was computed to reduce fill. For each sparse nonsymmetric problem, a minimum degree reordering with multiple elimination was used [18]. In all tables, n is the order of the matrix, and the second column ("sequential") contains the execution times in seconds required by the sequential algorithm.

Like most parallel algorithms, we see from the tables that, for a fixed number of processors, the efficiency increases as the size of the problem increases. However, for a fixed problem, the efficiency decreases as the number of processors increases. The efficiencies on the Intel/iPSC-2 are poor; this is mainly because the communication overhead is relatively high compared to the speed of computation.

Acknowledgments

Research was supported by the Applied Mathematical Sciences Research Program of the Office on Energy Research, U.S. Department of Energy.

References

[1] Alaghband, G. and H. F. Jordan, *Multiprocessor Sparse L/U Decomposition with Controlled Fill-in*. Technical Report 85-48, ICASE, NASA Langley Research Center, Hampton, Virginia, 1985.

[2] Chu, E.C.H., J. A. George, J. W-H. Liu, and E. G-Y. Ng, *User's Guide for SPARSPAK-A: Waterloo Sparse Linear Equations Package*, Technical Report CS-84-36, Dept. of Computer Science, University of Waterloo, Waterloo, Ontario, 1984.

[3] Duff, I.S., *MA28 - A Set of FORTRAN Subroutines for Sparse Unsymmetric Linear Equations*, Technical Report AERE R-8730, Harwell, 1977.

[4] Duff, I.S., "Parallel Implementation of Multifrontal Schemes," *Parallel Computing*, 3, 193–204, 1986.

[5] Duff, I.S., A. M. Erisman, and J. K. Reid, *Direct Methods for Sparse Matrices*, Oxford University Press, Oxford, England, 1987.

[6] Duff, I.S., N. I. M. Gould, M. Lescrenier, and J. K. Reid, *The Multifrontal Method in a Parallel Environment*, Technical Report CSS 211, Computer Science and Systems Division, Harwell, 1987.

[7] Duff, I. S., and J. K. Reid. "The Multifrontal Solution of Indefinite Sparse Symmetric Linear Equations," *ACM Trans. Math. Software*, 9, 302–325, 1983.

[8] Eisenstat, S. C., M. C. Gursky, M. H. Schultz, and A. H. Sherman, "The Yale Sparse Matrix Package: I. The Symmetric Codes," *Int. J. Num. Meth. Eng.*, 18, 1145–1151, 1982.

[9] Geist, G. A., and E. G-Y. Ng, *A Partitioning Strategy for Parallel Sparse Cholesky Factorization*, Technical Report ORNL/TM-10937, Oak Ridge National Laboratory, Oak Ridge, TN, 1988.

[10] George, J. A., "Nested Dissection of a Regular Finite Element Mesh," *SIAM J. Numer. Anal.*, 10, 345–363, 1973.

[11] George, J. A., M. T. Heath, and J. W-H. Liu, "Parallel Cholesky Factorization on a Shared-Memory Multiprocessor," *Linear Algebra and its Appl.*, 77, 165–187, 1986.

[12] George, J. A., M. T. Heath, J. W-H. Liu, and E. G-Y. Ng, "Solution of Sparse Positive Definite Systems on a Shared Memory Multiprocessor," *Internat. J. Parallel Programming*, 15, 309–325, 1986.

[13] George, J. A., M. T. Heath, J. W-H. Liu, and E. G-Y. Ng, "Symbolic Cholesky Factorization on a Local-Memory Multiprocessor," *Parallel Computing*, 5, 85–95, 1987.

[14] George, J. A., M. T. Heath, J. W-H. Liu, and E. G-Y. Ng, *Solution of Sparse Positive Sefinite Systems on a Hypercube*, Technical Report ORNL/TM-10865, Oak Ridge National Laboratory, Oak Ridge, TN, 1988.

[15] George, J. A., M. T. Heath, J. W-H. Liu, and E. G-Y. Ng, "Sparse Cholesky Factorization on a Local-Memory Multiprocessor. *SIAM J. Sci. Stat. Comput.*, 9, 327–340, 1988.

[16] George, J. A. and J. W-H. Liu, "An Automatic Nested Dissection Algorithm for Irregular Finite Element Problems," *SIAM J. Numer. Anal.*, 15, 1053–1069, 1978.

[17] George, J. A. and J. W-H. Liu, *Computer Solution of Large Sparse Positive Definite Systems*, Prentice-Hall Inc., Englewood Cliffs, New Jersey, 1981.

[18] George, J. A. and J. W-H. Liu, "On the Evolution of the Minimum Degree Algorithm," *SIAM Review*, to appear, 1989,

[19] George, J. A., J. W-H. Liu, and E. G-Y. Ng, "Communication Reduction in Parallel Sparse Cholesky Factorization on a Hypercube," in *Hypercube Multiprocessors*, M. T. Heath, (ed.), SIAM Publications, Philadephia, PA, 576–586, 1987.

[20] George, J. A., J. W-H. Liu, and E. G-Y. Ng, "A Data Structure for Sparse QR and LU Factors," *SIAM J. Sci. Stat. Comput.*, 9, 100–121, 1988.

[21] George, J. A., and E. G-Y. Ng, "Symbolic Factorization for Sparse Gaussian Elimination with Partial Pivoting," *SIAM J. Sci. Stat. Comput.*, 8, 877–898, 1987.

[22] George, J. A., and E. G-Y. Ng, *Parallel Sparse Gaussian Elimination with Partial Pivoting*, Technical Report ORNL/TM-10866, Oak Ridge National Laboratory, Oak Ridge, TN, 1988.

[23] Gilbert, J. R., *An Efficient Parallel Sparse Partial Pivoting Algorithm*, Technical Report CMI No. 88/45052-1, Centre for Computer Science,

Dept. of Science and Technology, Chr. Michelsen Institute, Bergen, Norway, 1988.

[24] Gilbert, J. R., and H. Hafsteinsson, *A Parallel Algorithm for Finding Fill in a Sparse Symmetric Matrix*, Technical Report TR 86-789, Dept. of Computer Science, Cornell University, Ithaca, New York, 1986.

[25] Gilbert, J. R., and T. Peierls, "Sparse Partial Pivoting in Time Proportional to Arithmetic Operations," *SIAM J. Sci. Stat. Comput.*, 9, 862–874, 1988.

[26] Gilbert, J. R., and E. Zmijewski, *A Parallel Graph Partitioning Algorithm for a Message-Passing Multiprocessor*, Technical Report TR 87-803, Dept. of Computer Science, Cornell University, Ithaca, New York, 1987.

[27] Gill, P. E., W. Murray, M. A. Saunders, and M. H. Wright, "Maintaining *LU* Factors of a General Sparse Matrix," *Linear Algebra and its Appl.*, 88/89, 239–270, 1987.

[28] Jess, J. A. G., and H. G. M. Kees, "A Data Structure for Parallel *L/U* Decomposition," *IEEE Trans. Comput.*, C-31, 231–239, 1982.

[29] Lewis, J. G. B.W. Peyton, and A. Pothen, *A Fast Algorithm for Reordering Sparse Matrices for Parallel Factorization*, Technical Report, Oak Ridge National Laboratory, Oak Ridge, TN, 1988. (Submitted to SIAM J. Stat. Sci. Comput.)

[30] Liu, J. W-H., *Reordering Sparse Matrices for Parallel Elimination*, Technical Report CS-87-01, Dept. of Computer Science, York University, Downsview, Ontario, 1987. (To appear in Parallel Computing.)

[31] Liu, J. W-H., *The Role of Elimination Trees in Sparse Factorization*, Technical Report CS-87-12, Dept. of Computer Science, York University, Downsview, Ontario, 1987. (To appear in SIAM J. Matrix Anal. & Appl.)

[32] Liu, J. W-H., "Equivalent Sparse Matrix Reordering by Elimination Tree Rotations," *SIAM J. Sci. Stat. Comput.*, 9, 424–444, 1988.

[33] Liu, J. W-H., and A. Mirzaian, *A Linear Reordering Algorithm for Parallel Pivoting: of Chordal Graphs*, Technical Report CS-87-02, Dept. of Computer Science, York University, Downsview, Ontario, 1987. (To appear in SIAM J. Disc. Math.)

[34] Pothen, A., *The Complexity of Optimal Elimination Trees*, Technical Report CS-88-16, Dept. of Computer Science, The Pennsylvania State University, University Park, PA, 1988.

[35] Sherman, A. H., *On the Efficient Solution of Sparse Systems of Linear and Nonlinear Equations*, Technical Report 46, Dept. of Computer Science, Yale University, New Haven, Connecticut, 1975.

[36] Sherman, A. H., "Algorithm 533 NSPIV, a FORTRAN Subroutine for Sparse Gaussian Elimination with Partial Pivoting," *ACM Trans. on Math. Software*, 4, 391–398, 1978.

[37] Wilkinson, J. H., *The Algebraic Eigenvalue Problem*, Oxford University Press, Oxford, 1965.

[38] Yannakakis, M., "Computing the Minimum Fill-In is NP-Complete," *SIAM J. Alg. Disc. Meth.*, 2, 77–79, 1981.

[39] Zmijewski, E., *Sparse Cholesky Factorization on a Multiprocessor*, PhD thesis, Department of Computer Science, Cornell University, Ithaca, New York 14853-7501, August 1987.

[40] Zmijewski, E., and J. R. Gilbert, "A Parallel Algorithm for Sparse Symbolic Cholesky Factorization on a Multiprocessor," *Parallel Computing*, 7, 199–210, 1988.

Chapter 12

Node Orderings and Concurrency in Structurally-Symmetric Sparse Problems

I. S. Duff* and S. L. Johnsson†

12.1 Introduction

The solution of structurally-symmetric sparse linear equations by direct methods offers an additional potential for concurrency over dense matrix problems. In this chapter we estimate the speed-up that is possible due to sparsity, and compare it with the concurrency that is inherent in the elimination within each pivot step. Johnsson [9] refers to these two forms of concurrency as concurrency in the elimination of multiple vertices (CEMV), and concurrency in the elimination of single vertices (CESV), although we will use alternative terms defined in Section 2 in this paper. Johnsson [7] gives analytical expressions for the attainable speed-up due to CEMV for some simple grid problems assuming unbounded parallelism. In this paper we also assume unbounded parallelism and, under this assumption, consider the potential speed-up both due to CEMV and CESV for some benchmark sparse matrix problems.

We represent the factorization in terms of an elimination tree in which each node represents the elimination operations corresponding to a single pivot step and the edges of the tree represent dependencies between different

---

*Computer Science and Systems Division, AERE Harwell, Didcot, Oxon, UK.
†Departments of Computer Science and Electrical Engineering, Yale University.

steps. This tree can be viewed as a computational graph for the solution process in the sense that operations at a node of the graph require information from the sons of the node. The tree thus defines a partial ordering; work corresponding to all leaf nodes can proceed immediately and concurrently and the sequential requirements are defined by the dependencies in the computational graph. The use of the elimination tree in implementing a parallel elimination scheme is discussed in Duff *et al.* [1]. Here we use the tree to study various measures of parallelism and compare some of our measures with those obtained on real systems. An elimination tree can be uniquely constructed for any given ordering. However, any ordering of the nodes of the tree which respects the dependencies is valid and involves the same number of floating-point operations.

The computations at each node include the assembly of small full matrices from the sons together with information on the pivot row and column corresponding to the node. The pivoting operations are then performed on the assembled small full matrix, called a frontal matrix, and the non-pivot rows and columns are sent to the father node. Because we are representing each Gaussian elimination step as a small full matrix, the rows and columns must be identified by integer index vectors and indirect addressing operations are required in the assembly phase.

Although we have defined the nodes of the elimination tree in terms of a single pivot, in practice we amalgamate nodes, so that a single tree node may correspond to several steps of Gaussian elimination. If the rows and columns in the son are a subset of those in the father then there is no loss in sparsity by amalgamating father and son pivots into the same tree node and performing both sets of eliminations at the single node. This node amalgamation can assist both vectorization and parallelism within the node. Indeed Duff and Reid [3] suggest performing additional node amalgamations at the cost of increasing slightly the overall work and storage. We do not perform this extra amalgamation here, but we do assume that each node of our tree may correspond to several steps of Gaussian elimination. Node amalgamation also facilitates numerical pivoting, since pivots can be chosen from anywhere within the rows and columns corresponding to the intended pivots from the original ordering. In the following, we do not assume diagonal dominance, and we include operation counts for pivoting operations.

We describe our model for calculating parallelism in Section 12.2 and the various orderings with which we experiment in Section 12.3. The results of our numerical experiments are given in Section 12.4. Finally, we present some concluding remarks in Section 12.5.

Algorithm 12.1 Sketch of tree-based Gaussian elimination.

```
DO over all nodes in an order respecting dependencies ⎫
    DO over number of pivots in node                  ⎬   (12.1)
                                                      ⎭

        Select pivot
        Compute multipliers
        For each non-pivot row in frontal matrix          (12.2)
            Update entries in rows                         (12.3)
```

12.2 Measures of Achievable Parallelism

In assessing the potential speed-up from different schemes for parallelizing sparse matrix factorization we only consider arithmetic and logic operations. Memory conflicts, bus or network delays, and the scheduling of a limited number of processing units are ignored. The reference measure we use for determining the available speed-up is simply the total amount of arithmetic for the sequential factorization, counting additions, multiplications, and divisions equally. We also include a count for the assembly of the node and for the selection of pivots within the frontal matrices since we do not assume diagonal dominance. A comparison is assumed to require the same time as an arithmetic operation. For each node ordering we use the elimination tree to define measures of parallelism.

To assist in describing our different measures of parallelism we sketch the Gaussian elimination procedure in Algorithm 12.1, where we have numbered the three basic loops in the elimination, noting that the overall number of executions of the two DOs in (12.2) is precisely the order of the matrix (the total number of pivots).

The first measure of parallelism is obtained only from the sparsity of the problem (corresponding to loop (12.2) in Algorithm 12.1). Here we assume that the operations at any one node must be done sequentially. The measure of parallelism is obtained by dividing the total sequential operation count by the maximum number of arithmetic operations, pivoting operations, and assembly operations along any path from a leaf node to the root of the tree. We call such a path for which a maximum count is obtained a "longest path" for that particular count. Since this parallelism is at the outermost level of the triply nested Gaussian elimination loop, we call this measure "outer" in our numerical experiments in Section 12.4.

The next two measures of parallelism are obtained from allowing parallelism within the elimination and assembly operations at a node. The two measures correspond to different granularity of operations. At the fine level of granularity in node elimination, a parallel algorithm is used for pivot selection, requiring a number of comparisons logarithmic in the size of the frontal matrix at the time the pivot is selected. The calculation of the multipliers (scaling) and the rank-1 update are performed concurrently. Parallelizing the assembly process is considerably simplified if the programming language includes a concurrent write instruction with addition, assuming that the architecture supports such instructions. The Ultracomputer (Schwartz [12]), the RP3 (Pfister *et al.* [10]), the Connection Machine (Hillis [6]) and the Fluent (Ranade *et al.* [11]) are examples of architectures that offer such support. We assume this model and parallelize the assembly accordingly. We call this measure "2 inner".

In architectures with a significant overhead for elementary operations such a fine level of granularity of operations may be infeasible, and an aggregation of operations desirable. For this reason we also compute a measure of parallelism for sequential pivot selection, followed by concurrent scaling and computation of all operations in the same matrix row for a single variable elimination. Different rows are treated sequentially. Hence, all columns of a frontal matrix are processed concurrently for the rank-1 update, but the operations in a column are performed sequentially (loop (12.3) only in Algorithm 12.1). The assembly is also performed sequentially within a column, but concurrently within a row. We call this second measure of parallelism for a node "1 inner" in Section 12.4.

Assuming that sufficiently many processors are available, different nodes of the elimination tree can be eliminated concurrently, as long as the dependency rules are not violated, in combination with concurrent elimination within individual nodes. We compute the parallelism that can be achieved by combining concurrent elimination of tree nodes, "outer", with the concurrent rank-1 update on rows and sequential column update, "1 inner", and call this measure "outer + 1 inner". We also determine the maximum parallelism attainable by concurrent elimination of tree nodes and maximally exploiting concurrency in each node elimination. We call this measure "outer+2 inner".

Below we give the precise definitions of how the different complexity measures are computed, and the corresponding measures of parallelism. The total number of pivots is n; that is, the entire sparse matrix is of size $n \times n$. The number of nodes in the elimination tree is N_T, and the number of pivots in node j is P_j. The size of the frontal matrix at node j is $K_j \times K_j$. The paths from the leaf nodes to the root are identified by $path_l$, where a unique l is associated with each leaf node. The number of additions for assembly of son i at node j is denoted $AA_j(i)$, and the number of sons of node j is S_j.

In the complexity expressions below, the term within the square brackets defines the number of operations required for one node of the elimination tree under the different schemes for exploiting parallelism. The first term in the complexity expression for a node represents the number of additions in sequence that are necessary for the assembly, the second term the number of comparisons in sequence for the pivot selection, the third term the number of arithmetic operations for the scaling, and the last term the number of sequential operations for the rank-1 update.

The complexity estimates for the different modes of parallelism are:

1. **Sequential:**

$$C_s = \sum_{j=1}^{N_T} \left[\sum_{i=1}^{S_j} AA_j(i) + \sum_{i=1}^{P_j} (K_j - i + 1) + \sum_{i=1}^{P_j} (K_j - l) + 2 \sum_{i=1}^{P_j} (K_j - l)^2 \right]$$

$$C_s = \sum_{j=1}^{N_T} \left[\sum_{i=1}^{S_j} AA_j(i) + P_j \left(1 + 2K_j \, (K_j + 1) - \frac{2}{3} (P_j + 1)(3K_j - P_j + 1) \right) \right]$$

2. **Outer:**

$$C_0 = \max_l \sum_{j=path_l}$$

$$\times \left[\sum_{i=1}^{S_j} AA_j(i) + \sum_{i=1}^{P_j} (K_j - i + 1) + \sum_{i=1}^{P_j} (K_j - i) + 2 \sum_{i=1}^{P_j} (K_j - i)^2 \right]$$

$$C_0 = \max_l \sum_{j=path_l}^{N_T} \sum_{j=1}$$

$$\times \left[\sum_{i=1}^{S_j} AA_j(i) + P_j \left(1 + 2K_j \, (K_j + 1) - \frac{2}{3} (P_j + 1) \, (3K_j - P_j + 1) \right) \right]$$

3. **1 inner:**

$$C_1 = \sum_{j=1}^{N_T} \left[\sum_{i=1}^{S_j} \sqrt{AA_j(i)} + \sum_{i=1}^{P_j} (K_j - i + 1) + \sum_{i=1}^{P_j} (K_j - i) + 2 \sum_{i=1}^{P_j} (K_j - i)^2 \right]$$

$$C_1 = \sum_{j=1}^{N_T} \left[\sum_{i=1}^{S_j} \sqrt{AA_j(i)} \frac{1}{2} P_j \, (6K_j - 3P_j + 1) \right]$$

4. 2 inner:

$$C_2 = \sum_{j=1}^{N_T} \left[\lceil \log_2 S_j \rceil + \sum_{i=1}^{P_j} \lceil \log_2 (K_j - i + 1) \rceil + 3P_j \right]$$

5. Outer + 1 inner:

$$C_{01} = \max_i \sum_{j \in path_i} \left[\sum_{i=1}^{S_j} \sqrt{AA_j(i)} + \sum_{i=1}^{P_j} (K_j - i + 1) + P_j + 2\sum_{i=1}^{P_j} (K_j - i) \right]$$

$$C_{01} = \max_i \sum_{j \in path_i} \left[\sum_{i=1}^{S_j} \sqrt{AA_j(i)} + \frac{1}{2} P_j (6K_j - 3P_j + 1) \right]$$

6. Outer + 2 inner:

$$C_{01} = \max_i \sum_{j \in path_i} \left[\lceil \log_2 S_j \rceil + \sum_{i=1}^{P_j} \lceil \log_2 (K_j - i + 1) \rceil + 3P_j \right]$$

Note that in determining the complexity measures C_o, C_{o1}, and C_{o2} the maximum is sought for different expressions, and the selected paths may be different. The paths with the widest frontal matrices are likely to benefit the most from parallelizing one or both "inner" loops. The path with the largest number of elimination operations may not be the same as the path with the maximum number of elimination operations given parallelization scheme "1 inner" or "2 inner". These two schemes may even yield different paths with respect to the maximization operation.

When deciding how to measure speed-ups in the parallel implementation of an algorithm, two base measures are used. One is the performance of the same algorithm on one processor, and the other is the time for the best known sequential algorithm for performing the same computation. The latter measure is what is important in practice. The former provides insight into how a given ordering performs when the number of processors is increased. We use this measure in most of our tables, but the latter measure is used occasionally for reference.

12.3 Orderings

Ordering strategies that attempt to minimize storage or the total amount of arithmetic performed during elimination have been investigated extensively. For general graphs the two most common ordering strategies are minimum degree and nested dissection. Nested dissection is of optimal order with respect to fill-in for graphs with good separators, such as planar graphs. However, the minimum degree ordering yields competitive orderings for many graphs and is often better than nested dissection at reducing fill-in and arithmetic (see, for example, Duff, Erisman, and Reid [1]).

An ordering that is optimal with respect to storage or arithmetic is not necessarily optimal with respect to parallel complexity. With unbounded parallelism, an ordering that minimizes the maximum number of pivots along any path in the tree is optimal with respect to solution time. If each node represents a single pivot step, then an ordering that minimizes the tree height (that is the number of nodes on the longest path from a leaf node to the root) is optimal, with respect to solution time, for unbounded parallelism.

A good example of the fact that an ordering that is optimal with respect to fill-in is far from optimal for any reasonable model of parallel computation is the solution of tridiagonal systems. For such systems a no fill-in ordering exists, but this ordering yields an inherently sequential algorithm. Cyclic reduction or, equivalently for this case, nested dissection yield a balanced tree with only $log_2 n$ levels and ample potential for parallelism for the solution of a system of order n. However, this ordering causes fill-in and approximately doubles the number of floating-point operations.

The cyclic reduction algorithm eliminates the maximum number of variables at each stage. However, such algorithms are in general not optimal (Johnsson [7]) with respect to minimizing the solution time even if the parallelism is unbounded. Furthermore, as will be demonstrated in Section 12.4, an ordering with a lower maximum of pivot steps (along any path from a leaf node to the root of the elimination tree) than another ordering does not necessarily have the lowest maximum number of arithmetic operations along any path in the two elimination trees (and certainly not the total amount of arithmetic, as is the case for cyclic reduction).

In addition to the minimum degree and nested dissection orderings, we also consider orderings that are specifically designed to enhance the parallelism in the tree. Since the height of the tree strongly affects the amount of sequential computation required, we try an ordering attempting to minimize the tree height. For this ordering we represent the matrix by a supervariable graph where each vertex can represent one or more variables. Pivots corresponding to variables in a supervariable will be eliminated at the same node of the elimination tree with amalgamated nodes. The ordering procedure starts by setting the depth of all vertices in the supervariable graph to 1, selecting one of these vertices (using minimum degree as a tie-breaker), and assigning a depth of two to all vertices to which this vertex is connected. We continue choosing vertices of depth 1 until all are exhausted and then search vertices of depth 2 and so on. Amalgamation is performed whenever possible. In each case, uneliminated vertices adjacent to vertices being eliminated are assigned a depth equal to the maximum of their current depth and one more than the vertex being eliminated. As in the case of the minimum degree ordering, the heuristic just described is local and will not necessarily minimize the tree height over all orderings, although, as we see from our results in Section 12.4, it usually does a very effective job.

The use of the elimination tree height as a measure of parallelism is rather crude, since the assumption is that all nodes are equal. A simple extension is to account for amalgamation and weight a node by its number of pivots. The ordering proceeds by setting the depth of a vertex to the maximum of its current depth and the depth of the vertex being eliminated plus the number of variables in the supervariable at that vertex. We study the effect of this modification in Section 12.4.

On the regular grid problems from a five-point discretization of the Laplacian operator, using one step of elimination according to a red-black ordering followed by, for instance, a minimum degree or a nested dissection ordering may yield a better result than either ordering alone, because we will have maximized the number of leaf nodes in the elimination tree. We also include this one-step "greedy" ordering for the grid problems. Note that the first step of minimum depth ordering may give a red-black ordering depending on the order selected by the minimum degree tie-breaker.

12.4   Results

The (partial) orderings of the variable eliminations in the sample problems have been obtained using general ordering routines. The minimum degree ordering routines are from the Harwell MA37 package (Duff and Reid [3]) and the nested dissection routines from Sparspak (George and Ng [5]). Our minimum height heuristic was obtained from a minor modification to the minimum degree code of MA37. The sample problems are chosen to represent both regularly structured (five-point discretization of the Laplacian on a two-dimensional grid) and less structured examples from the Harwell-Boeing test collection (Duff, Grimes, and Lewis [2]).

For $N \times N$ and $N \times N \times N$ grid problems the number of arithmetic operations at the root of the elimination tree is of the same order as the total number of operations. It follows that the parallelism that can be achieved from concurrent elimination of different pivots is $O(1)$. Thus exploiting sparsity alone for parallelism may yield very limited speed-ups. But the speed-up from exploiting concurrency in one inner loop is $O(N)$ for a two dimensional problem and $O(N^2)$ in three dimensions. If concurrency is obtained over both inner

Table 12.1: Speed-ups due to various levels of parallelism on our ideal machine. Minimum degree ordering used throughout.

	30 × 30	10 × 10	LUNDA	ERIS1176	BCSSTK24
Outer	2.4	2.1	1.9	1.3	2.7
Outer + 1 inner	46	18	25	40	310
1 inner	12	6	12	20	81
2 inner	65	25	80	88	2093
Outer + 2 inner	421	104	173	468	12375

Table 12.2: Statistics for 30 × 30 grid problem.

Ordering	MD	ND	GMD1	GMD2	RB-MD	RB-ND
Height	24	13	11	11	21	12
Total ops (×10^3)	415	514	894	1046	472	457
Counts on longest path						
Number of pivots	124	80	118	112	119	76
Number of operations (×10^3)	172	129	415	604	194	123
Row ops (×10^3)	9	6	14	16	9	6
Elim ops	985	643	1008	977	951	613

loops the speed-ups are $O(N^2/\log_2 N)$ and $O(N^4/\log_2 N)$ in two and three dimensions, respectively. For a problem of order n that can be represented as a banded matrix with bandwidth $m \ll n$ the speed-up is $O\left(\frac{n/m}{\log_2(n/m)}\right)$ from exploiting concurrency in the outer loop, while one inner loop yields a speed-up of $O(m)$ and both inner loops a speed-up of $O(m^2/\log_2 m)$. The results in Table 12.1 confirm that for the test cases only a small speed-up can be obtained from sparsity alone. When combined with sparsity within the nodes a more encouraging speed-up is obtained, as expected. The speed-up for one inner loop is always greater than the relative speed-up from parallelizing also the second loop, and sometimes significantly so. The concurrency in sparse elimination is substantial even in problems of small to moderate size, if all three loop levels are parallelized. Although the full "outer + 2 inner" figure may be difficult to obtain on an actual machine, the target can be viewed as an attractive goal for machines designed to work well at that fine a granularity.

In Table 12.2, the orderings are designated by MD (minimum degree), ND (nested dissection), GMD1 and GMD2 for the minimum height algo-

Table 12.3: Speed-up for 30 × 30 grid problem.

Ordering	MD	ND	GMD1	GMD2	RB-MD	RB-ND
Outer	2.4	4.0	2.1	1.7	2.4	3.4
Outer + 1 inner	46	80	66	67	50	75
1 inner	12	13	20	23	13	13
2 inner	65	79	138	161	74	71
Outer + 2 inner	421	799	887	1070	496	746
Rel. to best ordering						
Outer + 1 inner	46	69	29	26	46	69

rithms using tree height and weighted tree height respectively, and RB-MD and RB-ND for red-black followed by minimum degree and nested dissection, respectively. The good performance of minimum degree as a sequential ordering for reducing the operation count is illustrated in Table 12.2, where it gives the lowest total operation count among the orderings used. The operations count includes arithmetic operations for assembly, pivoting, and elimination. The numer of operations in sequence for parallel assembly, parallel pivoting, and parallel elimination in the case of parallelization by "outer + 2 inner" is labeled Elim ops in the table. The nested dissection ordering has a fairly good performance as a sequential ordering and its power as an ordering scheme for parallel elimination is seen in the last three rows of Table 12.2. The combination with a red-black ordering gives some improvement in the parallel orderings, although RB-MD is worse as a sequential ordering than minimum degree by itself. The two minimum height orderings reduce the height well, but even GMD2 is not as good as nested dissection at reducing the number of pivots on the longest path. Indeed GMD2 shows no real advance over GMD1 although it yields slightly fewer operations at the finest level of granularity. For this regular grid problem neither of them are as successful as nested dissection (or RB-ND) for exploiting parallelism.

Most of the speed-ups in Table 12.3 are relative to the sequential operations for the same ordering. The speed-up relative to the best sequential ordering is included for the "outer + 1 inner" ordering in the last row of the table, where we show the speed-ups relative to the total number of operations for the minimum degree ordering, the best sequential ordering. The better parallelization properties of nested dissection mean that it may be the best algorithm on a parallel machine, even if the number of sequential operations is higher than for the minimum degree ordering. This is evident from the last row of Table 12.3. It is clear that nested dissection (or RB-ND) is the best parallel ordering in spite of its poorer sequential performance.

Table 12.4: Statistics for ERIS1176 matrix from test collection.

Ordering		MD	ND	GMD1	GMD2
Height		50	66	12	17
Total ops	$(\times 10^3)$	618	873	645	636
Counts on longest path					
Number of pivots		175	111	103	82
Number of operations	$(\times 10^3)$	478	362	356	349
Row ops	$(\times 10^3)$	16	12	11	10
Elim ops		1319	947	841	692

Table 12.5: Speed-up for ERIS1176 problem.

Ordering	MD	ND	GMD1	GMD2
Outer	1.3	2.4	1.8	1.8
Outer + 1 inner	40	75	61	65
1 inner	20	21	20	20
2 inner	88	118	92	91
Outer + 2 inner	468	922	767	912
Relative to best ordering				
Outer + 1 inner	40	52	56	62

The less regular problem arises in circuit analysis, and the results in Tables 12.4 and 12.5 show a different performance to that for the regular grid problem. The minimum degree ordering is still the best sequential ordering, but nested dissection is now much poorer as a sequential ordering although it still beats minimum degree at exploiting sparsity. However, the minimum height algorithms are more effective here at reducing tree height and the number of pivots on the longest path; GMD2 is the best ordering for the parallel operation counts, in spite of producing a tree with greater height than GMD1.

12.5 Conclusions

In a multiprocessor architecture with few processing elements relative to the number of variables in the problem an "optimum" ordering is likely to be "close" to an ordering that is desirable for a sequential machine, even if communication complexity is included. For an architecture in which the number of processing elements is of an order comparable to the number of variables, the parallelism with respect to concurrent elimination of different pivots is bounded only for the first few levels of the elimination tree. With the number of processing elements falling in the intermediate range there is a choice of exploiting various combinations.

Our results in Section 12.4 confirm that the nested dissection ordering is very effective for exploiting parallelism in regular grid problems, but our new orderings based on minimizing the tree height show great promise for less regular problems.

In conclusion, there is much scope for exploiting concurrency in sparse elimination for any ordering, although it is important to utilize the parallelism both within the elimination tree nodes as well as across the nodes. If fine granularity working, at the level of a couple of arithmetic operations, is feasible, extremely high parallelism is available.

References

[1] Duff, I. S., A. M. Erisman, and J. K. Reid, *Direct Methods for Sparse Matrices,* Oxford University Press, London, 1986.

[2] Duff, I. S., R. G. Grimes, and J. G. Lewis, *Sparse Matrix Test Problems.* Report CSS 191, Computer Science and Systems Division, Harwell Laboratory, *ACM Trans. Math. Softw.* 15, 1, 1–14, 1989.

[3] Duff, I. S. and J. K. Reid, "The Multifrontal Solution of Indefinite Sparse Symmetric Linear Equations," *ACM Trans. Math. Softw.,* 9, 302–325, 1983.

[4] Duff, I. S. and J. K. Reid, "The Multifrontal Solution of Unsymmetric Sets of Linear Systems," *SIAM J. Sci. Stat. Comput.,* 5, 633–641, 1984.

[5] George, A. and E. Ng, *SPARSPAK: Waterloo Sparse Matrix Package User's Guide for SPARSPAK-B,* CS-84-37, Department of Computer Science, University of Waterloo, Ontario, Canada, 1984.

[6] Hillis, W. D., *The Connection Machine,* MIT Press, Cambridge, Massachusetts, 1985.

[7] Johnsson, S. L., *A Note on Householder's Method, Sparse Matrices and Concurrency,* Memo 4089, Department of Computer Science, California Institute of Technology, 1980.

[8] Johnsson, S. L., "Cyclic Reduction on a Binary Tree," *Computer Physics Communications,* 37, 195–203, 1985.

[9] Johnsson, S. L., *Fast Banded Systems Solvers for Ensemble Architectures,* Report YALEU/DCS/RR-379, Department of Computer Science, Yale University, Connecticut, 1985.

[10] Pfister, G. F., W. C. Brantley, D. A. George, S. L. Harvey, W. J. Kleinfelder, K. P. McAuliffe, E. A. Melton, V. A. Norton, and J. Weiss, "The IBM Research Parallel Processor Prototype (RP3): Introduction and Architecture," in *Proceedings of the 1985 International Conference on Parallel Processing, IEEE Computer Society,* 764-771, 1985.

[11] Ranade, A. G., S. N. Bhatt, and S. L. Johnsson, "The Fluent Abstract Machine," in *Advanced Research in VLSI, Proceedings of the Fifth MIT VLSI Conference,* MIT Press, Cambridge, Massachusetts, 71-93, 1988.

[12] Schwartz, J. T., "Ultracomputers," *ACM Trans. Program. Lang. Syst.,* 2, 484-521, 1980.

Chapter 13

A Multilevel Solution Method for Nine-Point Difference Approximations

*O. Axelsson**

13.1 Introduction

Consider the Poisson equation, $-\Delta u = f$, on a domain Ω consisting of a union of rectangles with sides parallel to the respective axes. Boundary conditions of Dirichlet, Neumann or periodic type are assumed with a consistent source function f in the case of pure Neumann or periodic boundary conditions.

The fourth-order nine-point difference approximation for this equation takes the form

$$- \Delta_h^{(9)} u_{i,j} = \tilde{f}_{i,j}^{(h)} \equiv f_{i,j} + \frac{h^2}{12} \Delta_h^{(5,+)} f_{i,j}, \ i,j \in \Omega_h , \qquad (13.1)$$

where Ω_h is the set of interior mesh points of the difference mesh, $u_{i,j}, f_{i,j}$ denote the values of u, f at a point i,j and $\Delta_h^{(5,+)}$ is the standard 5-point difference approximation. For later use we remark here that $\Delta_h^{(5,\times)}$ denotes the cross direction 5-point difference approximation to Δ. The difference

*Department of Mathematics, University of Nijmegen, The Netherlands.

stencil corresponding to (13.1) is

$$\left(-\Delta_h^{(9)}\right) = \frac{1}{6h^2}\begin{bmatrix} -1 & -4 & -1 \\ -4 & 20 & -4 \\ -1 & -4 & -1 \end{bmatrix}$$

$$= \frac{2}{3h^2}\begin{bmatrix} 0 & -1 & 0 \\ -1 & 4 & -1 \\ 0 & -1 & 0 \end{bmatrix} + \frac{1}{6h^2}\begin{bmatrix} -1 & 0 & -1 \\ 0 & 4 & 0 \\ -1 & 0 & -1 \end{bmatrix}$$

and is hence of positive type. By standard methods, the discretization error can be bounded by the truncation error times a constant $\left(\left\|\Delta_h^{(9)-1}\right\|\right)$, which is independent of h. Therefore, the nine-point difference approximation (13.1) has a discretization error $O(h^4)$, if $u \in C^{(6)}(\Omega)$, and even $O(h^6)$ in special cases. This method can, therefore, be quite efficient to use in practice because one can get an accurate solution on correspondingly coarser meshes as compared to an $O(h^2)$ method.

However, even so, one needs a fast solution method. Multigrid methods (see Hackbusch [7], for instance) can be used, but for a rate of convergence of optimal order one must require that an elliptic regularity assumption, bounding the norm of the second derivatives with the norm of the data f, be satisfied (see Ciarlet [5] and Axelsson and Barker [1], for instance). This regularity is not satisfied for a problem with a re-entrant corner or even less so for a problem domain with a 'crack' as is illustrated in Figure 13.1. (For a detailed discussion of the regularity of solutions to elliptic problems, see Grisvard [6].) For such problems the convergence rate of the multigrid method can be significantly degraded.

In this presentation an iterative solution method is presented, which is based purely on algebraic properties and hence independent of the elliptic regularity. The computational complexity per mesh point is $O(\log h)$, and furthermore, the "hidden" constant factor in this asymptotic result can be estimated. The method exploits the multilevel structure of the problem that arises when we delete diagonalwise cross couplings in the 9-point difference mesh. Diagonal couplings at the odd points ($i + j$ odd) are removed to form a 5-point coupled difference approximation and the resulting mutually decoupled set of odd points eliminated (as is illustrated in Figure 13.2a). This yields a mesh with nine-point couplings oriented in diagonalwise directions.

Here the vertical and horizontal couplings are deleted at points where i and j are even, to form a cross-coupled 5-point difference approximation as is illustrated in Figure 13.2b. These points are, therefore, now mutually uncoupled and are eliminated to form a difference mesh with nine-point couplings

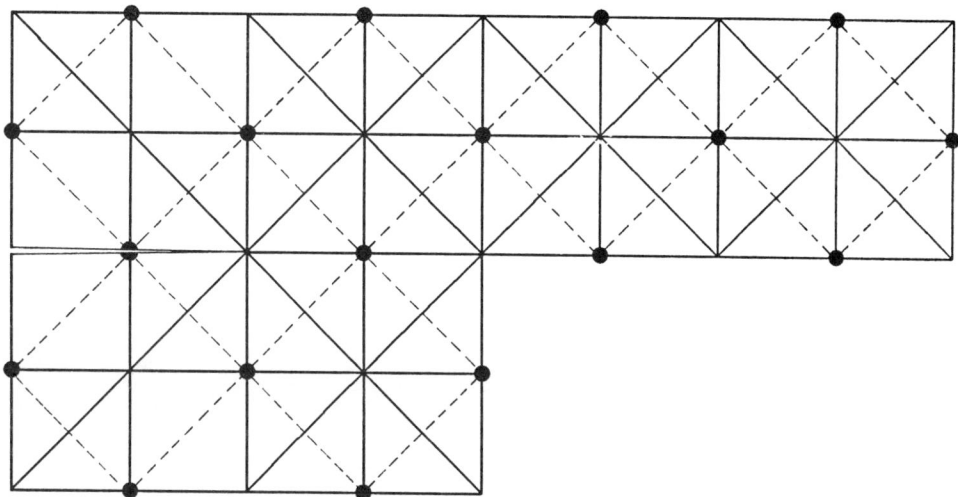

Figure 13.1: A nine-point difference mesh for a domain with a "crack" and a re-entrant corner. Dashed lines denote deleted couplings and black points (•) denote nodes to be eliminated on the current level.

in all points oriented as in the initial mesh (as in Figure 13.1) but for the double spaced mesh Ω_{2h} consisting of points where i and j are odd.

This method of constructing a sequence of 9-point – 5-point difference meshes can be continued until we arrive at a coarse mesh Ω_H, with $H = 2^{\ell/2}h$, where the sparse system order is so small that the most efficient solution method is a direct solution method.

In the method to be described we shall construct a preconditioner to the nine-point difference matrix which is spectrally equivalent to it. It turns out that it is most convenient to construct this preconditioner basing it on piecewise linear finite element matrices for isosceles right angled triangular meshes. Also it turns out that the relative condition number will be independent of the jumps in the (positive) coefficients a of the problem

$$- \nabla \cdot (a\nabla u) = f , \quad x \in \Omega \tag{13.2}$$

if a is constant on each element of the coarsest mesh Ω_H (seen as a triangular mesh). Our method is, therefore, applicable on this more general problem. The analysis requires then a so-called extended $C - B - S$ inequality. The preconditioner will be defined recursively using a matrix polynomial that approximates the Schur complements. This preconditioner contains the preconditioner on the next (coarser) mesh. Therefore, the preconditioner cannot be presented in explicit form. The implementation of the method requires only

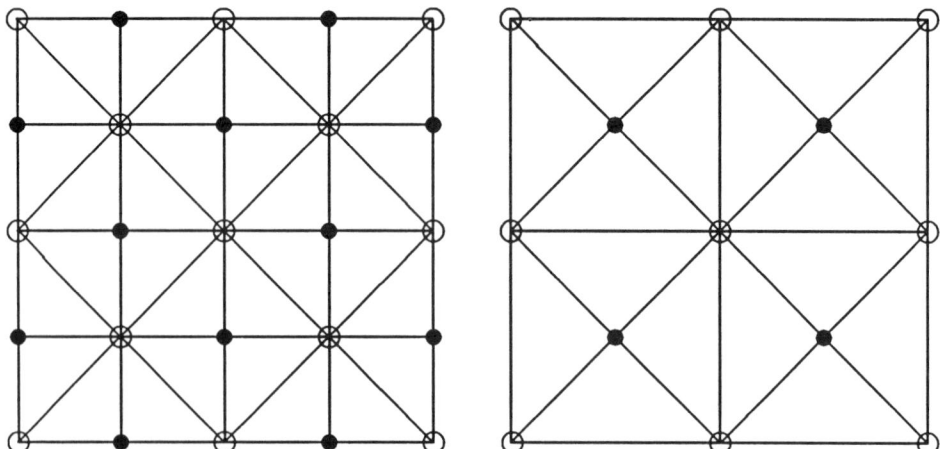

Figure 13.2: Two levels of meshes for 9-point–5-point differences (The mutually uncoupled black points are eliminated to form the next Schur complement.)

matrix vector multiplications and is, therefore, suitable for both parallel and vector computers.

Methods of the recursive, two-level type to be used in the present paper have previously been studied in more general contexts by Axelsson and Vassilevski [3,4]. A similar method to the present one has been studied by Kuznetsov [8], for the five-point difference matrix.

Given the matrix A, an initial vector x^0, the linear system $Ax = b$ corresponding to the nine-point difference approximation, and the preconditioner C, we can solve $Ax = b$ by the iterative method

$$C\delta^{n+1} = b - Ax^n ,$$

$$x^{n+1} = x^n + \frac{1}{\tau_n}\delta^{n+1}, \quad n = 0,1,\dots \qquad (13.3)$$

Here τ_n is a sequence of acceleration parameters computed as the roots of the shifted Chebyshev polynomial $T_k\left(\frac{2\lambda-b-a}{b-a}\right)$, where a, b are the extreme eigenvalues of $C^{-1}A$ or, more generally, lower and upper bounds of these eigenvalues, respectively, where $a > 0$.

Alternatively, we can use a preconditioned form of the conjugate gradient method to solve $Ax = b$. For an exposition of such methods, see for instance, Axelsson and Barker [1].

The remainder of the chapter is organized as follows. In Section 13.2 we define the sequence of combined 9-point – 5-point approximations basing them on a sequence of triangular meshes of the form illustrated in Figure 13.3. We also show some basic properties these matrices must satisfy. Section 13.3 contains the definition of the preconditioner and its efficient implementation. In Section 13.4 we derive the relative condition number of the preconditioner. The final section contains an estimate (upper bound) of the number of arithmetic operations per mesh point required by the method to reduce the error to a level ϵ, as well as a discussion of improvements and extensions to the method.

13.2 Two-Level Hierarchical Basis and Matrices

To describe the matrices of combined 9-point – 5-point form that will be used in the construction of the preconditioner of the 9 point difference matrix, we define first the two-level hierarchical basis functions. Consider then a mesh of vertical, horizontal and diagonal lines as shown in Figure 13.3, where at each vertex formed by solid lines we have defined piecewise linear basis functions with support on the triangles with solid line edges sharing this vertex. These basis functions form the set of old basis functions denoted

$$V_2^{(k)} = \left\{ \varphi_r^{(k)} \right\}, \quad r \in N_2^{(k)}$$

Consider next a refined mesh, obtained with the horizontal and vertical medians in the old mesh triangles, and indicated by dashed lines in Figure 13.3. The new set of basis functions corresponds to the set of midpoints $N_1^{(k+1)}$ on each edge on the horizontal and vertical meshlines of the old mesh indicated by circles in Figure 13.3. This set is denoted

$$V_1^{(k+1)} = \left\{ \varphi_r^{(k+1)} \right\}, \quad r \in N_1^{(k+1)}$$

and the complete set of basis functions on level $(k+1)$ is

$$V^{(k+1)} = V_1^{(k+1)} \cup V_2^{(k)}.$$

When we order the node points in the same fashion, taking first the new points and then the old points, the coefficient matrix takes the form of a two by two block matrix,

$$A^{(k+1)} = \begin{bmatrix} A_{11}^{(k+1)} & A_{12}^{(k+1)} \\ A_{21}^{(k+1)} & A_{22}^{(k+1)} \end{bmatrix} \begin{array}{l} \text{(new subset, } i+j \text{ odd)} \\ \text{(old subset, } i+j \text{ even)} \end{array} \qquad (13.4)$$

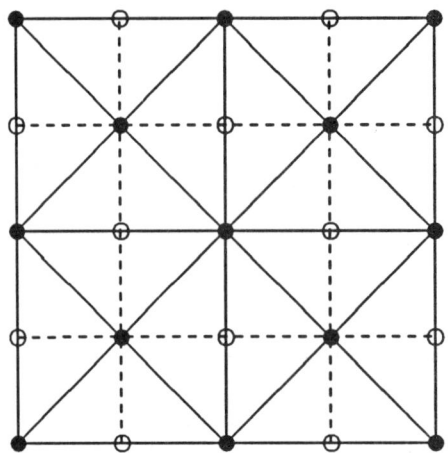

Figure 13.3: A 9-point – 5-point mesh for two-level hierarchical basis functions.

Here the entries are defined by

$$\left(A_{11}^{(k+1)}\right)_{rs} = B\left(\varphi_s^{(k+1)}, \varphi_r^{(k+1)}\right) , \quad r, s \in N_1^{(k+1)}$$

$$\left(A_{12}^{(k+1)}\right)_{rs} = B\left(\varphi_s^{(k)}, \varphi_r^{(k+1)}\right) , \quad r \in N_1^{(k+1)}, \ s \in N_2^{(k)}$$

$$\left(A_{21}^{(k+1)}\right)_{rs} = B\left(\varphi_s^{(k+1)}, \varphi_r^{(k)}\right) , \quad r \in N_2^{(k)}, \ s \in N_1^{(k+1)}$$

$$\left(A_{22}^{(k+1)}\right)_{rs} = B\left(\varphi_s^{(k)}, \varphi_r^{(k)}\right) , \quad r, s \in N_2^{(k)}$$

where the bilinear form $B(\cdot, \cdot)$ in the weak statement corresponding to (13.2) is defined by

$$B(u, v) = \iint_\Omega a \nabla u \cdot \nabla v \, d\Omega \tag{13.5}$$

(For notational simplicity, we have assumed here that the boundary conditions are homogeneous.) Here a is constant on each triangle of the coarsest mesh. On every level $(k + 1)$ we construct such two-level hierarchical bases and matrices. The matrix $A^{(k+1)}$ satisfies the following fundamental properties:

(i) $A_{11}^{(k+1)}$ is diagonal and its order is asymptotically the same as for the matrix $A^{(k)}$.

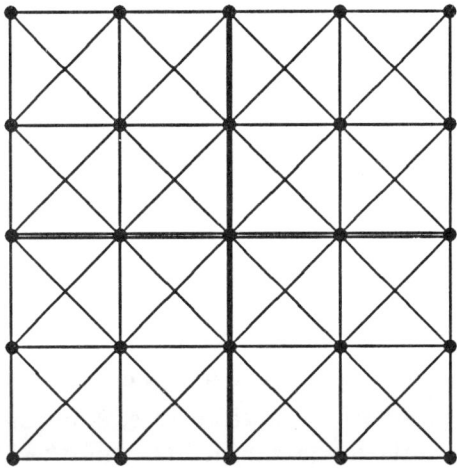

Figure 13.4: A difference mesh divided in four subdomains.

(ii) There exists a positive $\gamma < 1$ such that

$$|v_1^t A_{12}^{(k+1)} v_2| \leq \gamma_k \left\{ v_1^t A_{11}^{(k+1)} v_1 \right\}^{1/2} \left\{ v_2^t A_{22}^{(k+1)} v_2 \right\}^{1/2}$$

$$\forall v_1 \in \bar{V}_1^{(k+1)} \, , \quad v_2 \in \bar{V}_2^{(k)} \, , \quad \gamma_k \leq \gamma \qquad (13.6)$$

or equivalently,

$$1 - \gamma_k \leq \frac{v_1^t A^{(k+1)} v_1}{v_1^t A_{11}^{(k+1)} v_1 + v_2^t A_{22}^{(k+1)} v_2} \leq 1 + \gamma_k$$

$$\forall v_1 \in \bar{V}_1^{(k+1)} \, , \quad v_2 \in \bar{V}_2^{(k)}$$

where $\bar{V}_1^{(k+1)}$, and $\bar{V}_2^{(k)}$ denote the vector spaces of dimension equal to the dimension of the function spaces $V_1^{(k+1)}$ and $V_2^{(k)}$ (excluding basis functions at Dirichlet boundary condition points), respectively.

Using the bilinear form (13.5), (13.6) can equivalently be written

$$|B(u,v)| \leq \gamma_k B(u,u)^{1/2} B(v,v)^{1/2} \quad \forall \, u \in V_1^{(k+1)} \, , \quad v \in V_2^{(k)} \, .$$

The existence of $\gamma_k < 1$ follows from the extended *C-B-S* inequality and, as has been shown in Axelsson and Gustafsson [2] (see also Axelsson and Barker [1]), in the present case of isosceles right angled triangles one finds $\gamma^2 = 1/2$.

Since $A^{(k+1)}$ is symmetric and positive (semi) definite, it can be factored in the form

$$A^{(k+1)} = \begin{bmatrix} A_{11}^{(k+1)} & 0 \\ A_{21}^{(k+1)} & I \end{bmatrix} \begin{bmatrix} I & A_{11}^{(k+1)-1} A_{12}^{(k+1)} \\ 0 & S^{(k)} \end{bmatrix}$$

where

$$S^{(k)} = A_{22}^{(k+1)} - A_{21}^{(k+1)} A_{11}^{(k+1)-1} A_{12}^{(k+1)}$$

is the Schur complement (Gauss reduced system). It is readily seen that if $A^{(k+1)}$ is semidefinite then $A_{22}^{(k+1)}$ and $S^{(k)}$ are semidefinite and if $A^{(k+1)}$ is positive definite, they are both positive definite. Note also that $A_{22}^{(k+1)}$ is the matrix corresponding to the bilinear form $B(\cdot, \cdot)$ and the standard nodal basis functions $\varphi_r^{(k)} \in V_2^{(k)}$.

As has been shown in Axelsson and Gustafsson [2], the following relation between the quadratic forms of the Schur complement and matrix $A_{22}^{(k+1)}$ is valid:

Lemma 13.1

$$1 - \gamma_k^2 \le \frac{v_2^t S^{(k)} v_2}{v_2^t A_{22}^{(k+1)} v_2} \le 1 .$$

Furthermore, there exists a transformation matrix, J, taking the nodal vectors for the hierarchical basis function into the nodal vectors for the standard basis function, and

$$J^t A^{(k)} J = A_{22}^{(k+1)} .$$

It was shown in [3] that

$$S^{(k)} = \bar{S}^{(k)} \equiv \bar{A}_{22}^{(k+1)} - \bar{A}_{21}^{(k+1)} \bar{A}_{11}^{(k+1)-1} \bar{A}_{12}^{(k+1)}$$

which is the Schur complement for matrix $\bar{A}^{(k+1)}$ corresponding to the standard nodal basis function. Here partitioning corresponds to the partitioning of $A^{(k+1)}$; i.e.,

$$\bar{A}^{(k+1)} = \begin{bmatrix} \bar{A}_{11}^{(k+1)} & \bar{A}_{12}^{(k+1)} \\ \bar{A}_{21}^{(k+1)} & \bar{A}_{22}^{(k+1)} \end{bmatrix}$$

Note that $A_{11}^{(k+1)} = \bar{A}_{11}^{(k+1)}$ and $\bar{A}^{(k)} = A_{22}^{(k+1)}$.

13.3 Preconditioner

The preconditioners $M^{(k+1)}$ to the sequence $A^{(k+1)}$ will be defined recursively as follows. Let

$$M^{(k+1)} = \begin{bmatrix} A_{11}^{(k+1)} & 0 \\ A_{21}^{(k+1)} & I \end{bmatrix} \begin{bmatrix} I & A_{11}^{(k+1)-1} A_{12}^{(k+1)} \\ 0 & \tilde{A}^{(k)-1} \end{bmatrix} \tag{13.7}$$

where $\tilde{A}^{(k)}$ is a polynomial approximation to $S^{(k)-1}$ to be defined below. Note first that

$$M^{(k+1)} = A^{(k+1)} + \begin{bmatrix} 0 & 0 \\ 0 & \tilde{A}^{(k)-1} - S^{(k)} \end{bmatrix} \tag{13.8}$$

Following Axelsson and Vassilevski [3], we let

$$\tilde{A}^{(k)} = \left[I - P_\nu \left(M^{(k)-1} S^{(k)} \right) \right] S^{(k)-1} , \tag{13.9}$$

where P_ν is a polynomial of degree ν, satisfying $0 \le P_\nu(t) \le 1$, $0 < t < 1$ and normalized at the origin, $P_\nu(0) = 1$.

Here $M^{(k)}$ is the preconditioner of $A^{(k)}$ defined in the same way as $M^{(k+1)}$. Therefore, the preconditioners are only implicitly defined. (On the coarsest level, we let $M^{(1)} = A^{(1)}$, the standard coefficient matrix.) The implementation of the preconditioner must, therefore, use a recursive algorithm. To solve a linear system $M^{(k+1)} \begin{pmatrix} x \\ y \end{pmatrix} = \begin{pmatrix} f \\ g \end{pmatrix}$ with a preconditioner on any level $k > 1$, we perform the following steps.

Forward substitution: Solve

$$\begin{bmatrix} A_{11}^{(k+1)} & 0 \\ A_{21}^{(k+1)} & I \end{bmatrix} \begin{bmatrix} \xi \\ \eta \end{bmatrix} = \begin{bmatrix} f \\ g \end{bmatrix} ,$$

that is, compute

$$\xi = A_{11}^{(k+1)-1} f$$

and

$$\eta = g - A_{21}^{(k+1)} \xi$$

Backward substitution: Solve

$$\begin{bmatrix} I & A_{11}^{(k+1)-1} A_{12}^{(k+1)} \\ 0 & \tilde{A}^{(k)-1} \end{bmatrix} \begin{bmatrix} x \\ y \end{bmatrix} = \begin{bmatrix} \xi \\ \eta \end{bmatrix} ,$$

that is, compute

$$y = \tilde{A}^{(k)}\eta$$

and

$$x = \xi - A_{11}^{(k+1)^{-1}} A_{12}^{(k+1)} y$$

The crucial part is the efficient computation of $y = \tilde{A}^{(k)}\eta$. This will be performed in the following way. Let

$$Q_{\nu-1}(t) = \frac{1 - P_\nu(t)}{t} = q_0 + q_1 t + \cdots + q_{\nu-1} t^{\nu-1} \ .$$

Then (13.9) implies that

$$y = \tilde{A}^{(k)}\eta = Q_{\nu-1}\left(M^{(k)^{-1}} S^{(k)}\right) M^{(k)^{-1}}\eta \qquad (13.10)$$

Note that the matrix polynomial in (13.10) is symmetric and positive definite. The computation of y in (13.10) can be performed in ν iteration steps: Let $y^{(0)} = 0$; for $r = 1$ to ν solve $M^{(k)} y^{(r)} = q_{\nu-r}\eta + S^{(k)} y^{(r-1)}$. Then (13.10) shows that $y = y^{(\nu)}$.

The linear systems with $M^{(k)}$ are solved by forward and backward substitution steps as shown above for $M^{(k+1)}$ and this is repeated from one level to the next. Eventually we reach the coarsest level, where $M^{(1)} = A^{(1)}$ and this system can be solved by a direct solution method, for instance. (Actually nothing prevents us from continuing the recursion until the coarsest level just contains one point.)

Note that the only computations involved are matrix-vector multiplications. Furthermore, these can be performed on an element-by-element basis. If we have p parallel working processors at our disposal, we can divide the domain, Ω, in p parts Ω_i of about equal size (see Figure 13.4) and let each processor P_i compute all matrix vector multiplications on the subdomain Ω_i. The only communication required between the processors occurs via matrix-vector multiplications for matrix rows corresponding to points on the common boundaries of the subdomains. Since these have a dimension one order less than the dimension of the set of internal points, the computer time will not be dominated by communication times.

13.3.1   Computational Complexity

The computational complexity of the preconditioner per mesh point for the finest mesh of each iteration is readily found to be

$$13\left(1 + \tfrac{\nu}{2} + \cdots + \left(\tfrac{\nu}{2}\right)^{\ell-1}\right) + \left(\tfrac{\nu}{2}\right)^\ell C_H = 13\ell + \left(\tfrac{\nu}{2}\right)^\ell C_H \ , \qquad (13.11)$$

if $\nu = 2$, $h = 2^{-\ell/2}H$, where $\ell = \log_2 H/h$, and C_H is the computational complexity to solve the problem on the coarsest mesh. Hence if H is fixed the computational complexity is $O(\log h^{-1})$ per mesh point of the finest level mesh.

The number of iterations will depend, among other things, on the choice of the polynomial P_ν. Several choices were discussed in Axelsson and Vassilevski [3], but the most efficient choice is based on Chebyshev polynomials as will be defined in the next section. We shall show that for this choice the preconditioner $M^{(k+1)}$ is spectrally equivalent to $A^{(k+1)}$.

13.4   Relative Condition Number

It is seen from (13.8) and (13.9) that

$$v^t M^{(k+1)} v \;=\; v^t A^{(k+1)} v + v_2^t \left(\tilde{A}^{(k)-1} - S^{(k)} \right) v_2 \qquad (13.12)$$

$$= \; v^t A^{(k+1)} v + v_2^t S^{(k)} P_\nu \left(M^{(k)-1} S^{(k)} \right)$$

$$\left[I - P_\nu \left(M^{(k)-1} S^{(k)} \right) \right]^{-1} v_2$$

for all $v \in \bar{V}^{(k+1)}$, where $v = (v_1, v_2)$ has been partitioned in two block vectors consistent with the partitioning of $A^{(k+1)}$.

The definition of the polynomial now shows that

$$v^t M^{(k+1)} v \geq v^t A^{(k+1)} v \quad \forall v \in \bar{V}^{(k+1)} \qquad (13.13)$$

Further, since

$$\inf_v v^t A^{(k+1)} v = \inf_{v_2} v_2^t S^{(k)} v_2$$

it follows from (13.13) that

$$\sup_v \frac{v^t M^{(k+1)} v}{v^t A^{(k+1)} v} \;=$$

$$1 + \sup_{v_2} \frac{v_2^t S^{(k)} P_\nu \left(M^{(k)-1} S^{(k)} \right) \left[I - P_\nu \left(M^{(k)-1} S^{(k)} \right) \right]^{-1} v_2}{v_2^t S^{(k)} v_2} \qquad (13.14)$$

Definition 13.1 Let

$$\lambda_{k+1} = \sup_v \left\{ v^t M^{(k+1)} v / v^t A^{(k+1)} v \right\}$$

Lemma 13.2. Let $t_k = \sup_{v_2}\left\{v_2^t S^{(k)} v_2 / v_2^t M^{(k)} v_2\right\}$.

Then $(1 - \gamma_k^2)\lambda_k^{-1} \le t_k \le 1$.

Proof. We have

$$\frac{v_2^t S^{(k)} v_2}{v_2^t M^{(k)} v_2} = \frac{v_2^t S^{(k)} v_2}{v_2^t A^{(k)} v_2} \frac{v_2^t A^{(k)} v_2}{v_2^t M^{(k)} v_2} \tag{13.15}$$

Using the transformation matrix J, defined in Section 13.2, we find

$$\frac{v_2^t S^{(k)} v_2}{v_2^t A^{(k)} v_2} = \frac{\hat{v}_2^t \bar{S}^{(k)} \hat{v}_2}{\hat{v}_2^t \bar{A}^{(k)} \hat{v}_2} = \frac{\hat{v}_2^t S^{(k)} \hat{v}_2}{\hat{v}_2^t A_{22}^{(k)} \hat{v}_2}$$

where $v_2 = J\hat{v}_2$. Lemma 13.1 now shows that

$$1 - \gamma_k^2 \le \frac{v_2^t S^{(k)} v_2}{v_2^t A^{(k)} v_2} \le 1$$

and the statement follows from (13.15), (13.13) and definition 13.1. ■

The following bounds for the condition number λ_{k+1} now follow from (13.14).

Theorem 13.1. Let $M^{(k+1)}$ be defined by (13.7) and (13.9). Then

$$1 \le \frac{v^t M^{(k+1)} v}{v^t A^{(k+1)} v} \le \frac{1}{1 - P_\nu(t_k^*)} \qquad \forall v \in \bar{V}^{(k+1)}$$

where $P_\nu(t_k^*) = \max P_\nu(t)$, $t_k \le t \le 1$. ■

Next, let

$$\hat{\lambda}_{k+1} = \frac{1}{1 - P_\nu(\hat{t}_k^*)}, \tag{13.16}$$

where $P_\nu(\hat{t}_k^*) = \max P_\nu(t)$, $(1 - \gamma_k^2)\hat{\lambda}_k^{-1} \le t \le 1$. Then Lemma 13.2 shows that $\hat{\lambda}_{k+1}$ is a majorizing sequence for λ_{k+1}. We shall now choose the polynomial P_ν such that its maximum is smallest of all polynomials of degree ν on the interval $[\alpha_k, 1] = \left[(1 - \gamma_k^2)\hat{\lambda}_k^{-1}, 1\right]$, subject to the constraint $P_\nu(0) = 1$ and the inequality $0 \le P_\nu(t) < 1$.

It is readily seen that the solution is

$$P_\nu(t) = P_{\nu,\alpha_k}(t) = \frac{T_\nu\left(\frac{1+\alpha_k - 2t}{1-\alpha_k}\right) + 1}{T_\nu\left(\frac{1+\alpha_k}{1-\alpha_k}\right) + 1}, \tag{13.17}$$

where $T_\nu(x) = \frac{1}{2}\left\{\left[x + (x^2 - 1)^{1/2}\right]^\nu + \left[x - (x^2 - 1)^{1/2}\right]^\nu\right\}$ is the Chebyshev polynomial. From (13.6) it follows that (if $\gamma_k = \gamma$)

$$\hat{\lambda}_{k+1} = \frac{1}{1 - P_\nu(\alpha_k)}, \qquad \alpha_k = \left(1 - \gamma^2\right)\hat{\lambda}_k^{-1} \qquad (13.18)$$

We shall now show that there exists a positive α such that $\alpha \leq \alpha_k$. This implies that there also exists a $\hat{\lambda}$ such that $\hat{\lambda}_{k+1} \leq \hat{\lambda}$, where

$$\hat{\lambda} = \frac{1}{1 - P_\nu(\alpha)}$$

To this end, note that (13.18) implies

$$\alpha_{k+1} = \left(1 - \gamma^2\right)\hat{\lambda}_{k+1}^{-1} = \left(1 - \gamma^2\right)Q_{\nu-1}(\alpha_k)\alpha_k. \qquad (13.19)$$

Since $(1 - \gamma^2)Q_{\nu-1}(1) = (1 - \gamma^2)(1 - P_\nu(1)) \leq 1 - \gamma^2 < 1$ there exists $\alpha > 0$ such that $\alpha_k \geq \alpha$ if

$$\left(1 - \gamma^2\right)Q_{\nu-1}(0) > 1 \qquad (13.20)$$

For the choice (13.17), we have $Q_{\nu-1}(0) = \nu^2$ and since $\gamma^2 = \frac{1}{2}$, (13.20) is satisfied if $\nu \leq 2$. For $\nu = 2$, the recursion (13.19) takes the form

$$\alpha_{k+1} = \left(1 - \gamma^2\right)\left(\frac{2}{1 + \alpha_k}\right)^2 \alpha_k, \qquad k = 1, 2, \ldots, \ell - 1 \qquad (13.21)$$

where $\alpha_1 = 1 - \gamma^2$, $\left(\hat{\lambda}_1 = 1\right)$. Hence, for $\gamma^2 = \frac{1}{2}$,

$$\alpha_1 = \frac{1}{2}, \quad \alpha_2 = \frac{4}{9}, \quad \alpha_3 = \frac{72}{169}, \quad \cdots,$$

$$\alpha_k \to \alpha = \sqrt{2} - 1.$$

and

$$\hat{\lambda} = \frac{T_2\left(\frac{1+\alpha}{1-\alpha}\right) + 1}{T_2\left(\frac{1+\alpha}{1-\alpha}\right) - 1} = \left(\frac{1+\alpha}{2}\right)^2 \cdot \frac{1}{\alpha}$$

$$= \frac{1}{2}\left(\sqrt{2} + 1\right) \simeq 1.207.$$

Notice that there is little difference in the values α_k of the recursion, so we can take the limit value α for all levels without losing much efficiency. We collect these results in

Theorem 13.2. Let $M^{(k+1)}$ $k = 0, 1, \cdots, \ell - 1$ be defined by (13.7), (13.9) and (13.17) where $\alpha = \sqrt{2} - 1$. Then $M^{(\ell)}$ is spectrally equivalent to $A^{(\ell)}$ with bounds

$$v^t A^{(\ell)} v \leq v^t M^{(\ell)} v \leq \hat{\lambda} v^t A^{(\ell)} v \quad \forall v \in \bar{V}^{(\ell)}$$

where $\hat{\lambda} = \frac{1}{2}\left(\sqrt{2} + 1\right)$. ∎

13.4.1   Computational Complexity

The 9-point – 5-point matrix $A^{(\ell)}$ on the finest level will be used as a preconditioner to the 9-point difference matrix A. It is readily seen that $A^{(\ell)}$ is spectrally equivalent to A with spectral condition number $\frac{3}{2}$. Therefore, the preconditioner $M^{(\ell)}$ is spectrally equivalent to A with condition number $\frac{3}{2}\hat{\lambda}$. It follows that the rate of convergence of the Chebyshev iteration method (13.3) is determined by the reduction rate

$$\rho = \frac{1 - \sqrt{2/3\hat{\lambda}}}{1 + \sqrt{2/3\hat{\lambda}}} \simeq 0.14$$

The number of iterations required for a relative iteration error ϵ is not more than n, where $\rho^n \leq \epsilon$ so that $n = \log\frac{1}{\epsilon} / \log\frac{1}{\rho}$. Taking the cost to compute the residual in (13.3) and the cost of the preconditioner in (13.11) into account, we find that the total cost is approximately

$$n23\ell\,(1 + C_H/13\ell) \leq 20\ell \log_{10}\left(\tfrac{1}{\epsilon}\right)(1 + C_H/13\ell) \tag{13.22}$$

By use of the conjugate gradient method we can expect even fewer iterations but at a somewhat higher cost per iteration resulting from certain inner products.

13.5   Concluding Remarks

We have presented a method to compute the solution of the nine-point difference approximation in a computational complexity proportional to the number of levels of meshes, i.e., typically a cost $O(\log h^{-1})$ per meshpoint. This result is even true for the case of discontinuous diffusion coefficients, if these coefficients are constant on each triangle of the coarsest mesh. The method requires only matrix-vector multiplications and can, therefore, be implemented efficiently on vector and parallel processors, using domain decompositions for instance.

We have also given an upper bound (13.22) of the actual number of arithmetic operations (flops) required for a relative iteration error ϵ. This bound contains the factor ℓ equal to the number of levels. Since the discretization error is $O(h^4)$, we can expect in many problems that the number of levels needed will not be large. When the method is implemented on a vector computer it may even be efficient to stop at a mesh that is still relatively fine, provided it contains sufficiently many interior points relative to the number of boundary points.

References

[1] Axelsson, O., and V.A. Barker, *Finite Element Solution of Boundary Value Problems*, Academic Press, Orlando, FL, 1984.

[2] Axelsson, O., and I. Gustafsson, "Preconditioning and Two-Level Multigrid Methods of Arbitrary Degree of Approximations," *Math. Comp.* 40, 219–242, 1983.

[3] Axelsson, O., and P. Vassilevski, "Algebraic Multilevel Preconditioning Methods I," Report 8811, Department of Mathematics, Catholic University, Nijmegen, The Netherlands, *Numer. Math.*, to appear, 1988.

[4] Axelsson, O., and P. Vassilevski, "Algebraic Multilevel Preconditioning Methods II," Report 1988–15, Institute for Scientific Computation, University of Wyoming, Laramie, Wyoming, 1988.

[5] Ciarlet, P. A., *The Finite Element Method for Elliptic Problems*, North-Holland Publ., Amsterdam, 1978.

[6] Grisvard, P., *Elliptic Problems in Nonsmooth Domains*, Pitman Publ., Boston - London - Melbourne, 1985.

[7] Hackbusch, W., *Multigrid Methods and Applications*, Springer-Verlag, Berlin - New York, 1985.

[8] Kuznetsov, Yu A., "Multigrid Domain Decomposition Methods for Elliptic Problems," in *Proceedings of VIII International Conference on Computational Methods for Applied Science and Engineering*, 2, 1987, INRIA, France, 605–616, 1987.

Chapter 14

A Structural Analysis Algorithm for Massively Parallel Computers

Robert E. Benner, Gary R. Montry,*
and John L. Gustafson**

14.1   Introduction

We currently are developing new methods, application programs, and performance models for massively parallel systems (Gustafson *et al.* [7]). Here, massive parallelism refers to both general-purpose MIMD and SIMD systems with 1000 or more autonomous processors. The high performance we have attained for several applications on a massively parallel system leads us to examine the relationship between Amdahl's law (Amdahl [1]) and two other models of parallel performance. We note that it can be much easier to reach a high degree of parallelism than Amdahl's law implies.

We present a massively parallel algorithm that reaches high parallel efficiency (99.5%) and high sustained performance (132 MFLOPS with 64-bit arithmetic) on structural analysis problems. Results are shown as a function of both problem size and number of processors on a 1024 processor system. Fixed problem size results in a parallel speedup of 502. Fixed execution time results in a parallel speedup of 987. Fixed problem size per processor results in a parallel speedup of 1019.

*Sandia National Laboratories, Albuquerque, NM.

Issues encountered in measuring parallel speedup are presented in Section 14.2. Machine parameters specific to our hypercube system are summarized in Section 14.3. Methods and results for the structural analysis application on ensembles of up to 1024 processors are discussed in Section 14.4. Algorithms for both host and node processors are also given in Section 14.4. Advances in our understanding of parallel processing applied to structural analysis are summarized in Section 14.4. We note that parts of this material appeared in Gustafson *et al.* [7].

14.2   Performance Models

Parallel performance depends on both problem size and ensemble size. Three subsets of this domain have received attention in the parallel processing community: fixed problem size (Amdahl [1]), fixed problem size per processor (Gustafson [6]), and fixed execution time (Gustafson [5]).

14.2.1   Fixed-sized Model

Let P be the number of processors in the parallel ensemble, s be uniprocessor time spent on serial parts of the program, and p be uniprocessor time spent on parts that could execute in parallel. Amdahl's law (Amdahl [1]) states that

$$\text{Speedup} = (s+p)/(s+p/P)$$

$$= 1/(s+p/P) \ . \tag{14.1}$$

Here, we have normalized serial time $s + p = 1$. For $P = 1024$, this is a steep function of s near $s = 0$ (slope of about $-P^2$). (14.1) gives an asymptotic speedup of $1/s$ at large P.

Amdahl's model applies to the first subset, the fixed-sized speedup line (Fig. 14.1). On this line, problem size is held fixed and ensemble size is varied. Speedup is then calculated as the ratio of the run time on one processor to that on P processors. On large parallel ensembles fixing problem size can be a severe constraint, because the problem must run efficiently when variables occupy a small fraction of each memory.

14.2.2   Scaled Model

In practice, a scientific computing problem often scales with the available processing power, either memory or speed. Most users will expand a problem (more spatial variables, for example) to use expanding hardware re-sources. The scaled model assumes the problem scales with memory. As a first approximation, work that can be done in parallel varies linearly with the ensemble size for the structural analysis problem. Therefore, when we double the ensemble size, we double spatial variables in a physical simulation.

This model is, in a sense, the inverse of Amdahl's model. Instead of asking how fast a given serial program would run on a parallel processor, we ask how long a given parallel program would run on a serial processor. If we use s' and p' to represent serial and parallel time spent on the parallel system, then one processor takes time $s' + p'P$ to do the task. Our reasoning gives a different law (Gustafson [6]) from that of Amdahl:

$$\text{Scaled speedup} \quad = \quad (s' + p'P)/(s' + p')$$

$$= \quad P + (1 - P)s' . \tag{14.2}$$

Here, we have normalized parallel time, $s' + p' = 1$. This function is simply a line, of moderate slope $1 - P$. When speedup is measured by scaling the problem storage, scalar fraction s tends to shrink as more processors are used. Therefore, it is much easier to reach efficient parallel performance than is implied by Amdahl's model, and speedup as a function of P is not bounded by an asymptote.

On the scaled speedup line (Fig. 14.1) processor power (memory and speed) increases with the ensemble size. The ratio of computation to com-munication is higher for scaled problems. One can model the performance of scaled problems with a hypothetical processor node that has direct unipro-cessor access to all the real random-access memory of the machine. This hypothetical processor performance is equivalent (after adjustment by a fac-tor introduced below) to the ratio of measured MFLOPS.

14.2.3 Fixed-Time Model

We now consider a model intermediate to the fixed size and scaled models. In this model, the constant is the execution time, the time a user is willing to wait for an answer.

It is common in scientific problems for operation count to grow faster than the variable count. For example, N by N dense matrix multiplication takes order N^2 storage but order N^3 operations, so operation count varies as the 3/2 power of variable storage. Ideally, operation/second rate is proportional to P. Constant storage per processor (scaled speedup) means N^2 scales with P, but constant job time means $N^{3/2}$ scales with P. That is, the matrix size N should grow as $P^{2/3}$. On the log-log graph in Fig. 14.1, this appears as

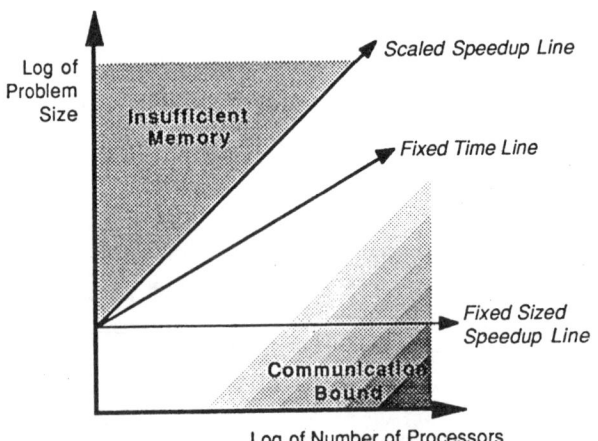

Figure 14.1: Ensemble Computing Performance Pattern.

a line of slope 2/3. It happens that the two-dimensional structural analysis problem presented here needs order N storage but order $N^{3/2}$ operations, so the 2/3 slope applies for it as well.

We note that fixed-time speedup, like fixed-size speedup, has an asymptote at large P. However, the asymptotic value of fixed-time speedup is quite large (about 10^{12}) for the structural analysis problem presented here.

14.2.4   Operation Efficiency

Converting an algorithm from serial to parallel often increases operation count. For example, it might be faster to have each processor calculate a globally needed quantity than to have one processor calculate and send it to other processors. We define $\Omega(N)$ as operation count for the best serial algorithm, where N measures problem size, and $\Omega_P(N)$ as operation count for the best parallel algorithm. Generally, $\Omega_P(N) \geq \Omega(N)$; less obviously, $\Omega_1(N) \geq \Omega(N)$, where $\Omega_1(N)$ is the operation count for the parallel algorithm executed on one processor. Hence, P processors can be 100% busy on computation and still be less than P times faster than the best serial algorithm. For example, suppose a two-dimensional simulation calls for a serial operation count given by

$$\Omega(N) = a + bN + cN^2 \quad . \tag{14.3}$$

Here, a, b, and c are nonnegative integers. In running the N by N problem in parallel on P processors, each processor treats a subdomain with operation cost $\Omega(N/\sqrt{P})$. This operation cost is more work than would be done by a serial processor:

$$\Omega_P(N) = P\Omega\left(\frac{N}{\sqrt{P}}\right)) \geq \Omega(N) \ . \tag{14.4}$$

Equality is possible if and only if $a = b = 0$ in (14.3).

We define an operation efficiency factor $\eta_P(N)$ by

$$\eta_P(N) = \frac{\Omega(N)}{\Omega_P(N)} \leq 1 \ . \tag{14.5}$$

Here, we let $\Omega_P(N)$ be a general function of P. When we refer to efficiency, we are accounting both for apparent efficiency (computation time divided by total time) and operation efficiency as defined in (14.5). Note that an exact floating point operation count is needed to make this measurement. Given operation efficiency and measured MFLOPS on one and P processors, $M(1)$ and $M(P)$ respectively, scaled speedup can be calculated as

$$\text{Scaled speedup} = \eta_P(N)\,M(P)/M(1) \ . \tag{14.6}$$

14.3   The NCUBE Parallel Computer

The NCUBE/ten is well suited for parallel speedup research. Each processor has enough resources (512K bytes, 0.13 MFLOPS) to run a problem of practical size (*e.g.*, 2048 bilinear or 1024 biquadratic finite elements) on a single processor. The 1024-node ensemble competes with conventional supercomputers on an absolute performance basis.

All memory is distributed in the hypercube architecture. Information is shared between processors by explicit communications across channels (as opposed to the shared memory approach of storing data in a common memory). An important parameter is the measured ratio of computation time to communication time in such a processor node. Currently, a 64-bit floating point operation takes about 7-8 μsec to execute on one node when using the Fortran compiler and indexed memory-to-memory operations. Our experience is that computationally intensive single-node Fortran programs fall within this range (0.12 to 0.14 MFLOPS). With system calls from Fortran, a message takes about 0.35 msec to start and then 2 μsec per byte; this time constrains speedup for a fixed-sized problem using 1024 processors. However, as discussed in Gustafson *et al.* [7], time to move data across a channel can be partly overlapped with computations or other communications.

An application with 4000 variables has only 4 variables per processor node when distributed over 1024 processors. For a conjugate gradient solver, each variable involves about 1000 floating point operations (about 12 msec) per iteration, for a total of 48 msec before data must be exchanged with neighbors. Data exchanges include four reads and four writes of 48 bytes each to nearest-neighbor processor, plus a global exchange of 24 bytes. This gives a worst-case time of $(4+4) \times (350+48 \times 2) + (10+10) \times (350+24 \times 2)$ μsec, or about 11.5 msec. Therefore, when a single-node problem is distributed on the entire 1024-processor ensemble, communication overhead will be about 20%. The cost of operation efficiency, about 16% in this case, is as severe as communication overhead for the structural analysis problem.

14.4 Structural Analysis Methods and Performance

14.4.1 Mathematical Formulation

The differential equations of equilibrium in plane elasticity used in this program are presented elsewhere (Szabo and Lee [12]) with their finite element formulation. They can be summarized as

$$\alpha u_{xx} + \beta v_{xy} + G(u_{yy} + v_{xy}) + F_x \;=\; 0 \;, \qquad (14.7)$$

$$\beta u_{xy} + \alpha v_{yy} + G(u_{xy} + v_{xx}) + F_y \;=\; 0 \;. \qquad (14.8)$$

Here, u and v represent displacement components in the x and y directions, F_x and F_y are force components, and α, β, and G are constitutive equation parameters. Subscripts on u and v denote partial derivatives.

A standard method for solving structural analysis problems involves using finite elements to approximate the physics and Preconditioned Conjugate Gradients (PCG) to solve the resulting large, sparse, linear system. These methods are used in both the solid mechanics application program JAC (Biffle [3]), a highly-vectorized production program for the CRAY X-MP, and the highly parallel program discussed herein. Jacobi (main diagonal) preconditioning is used in both programs because it vectorizes and can be made parallel. The program uses 64-bit precision throughout.

The parallel program never forms the stiffness matrix in memory, and hence can solve systems of equations several times larger than if the matrix were stored explicitly. The only place that the stiffness matrix appears in the standard PCG algorithm (e.g., Golub and Van Loan [4]) is the matrix-vector product with projection vector p. If residual and iterate vectors are denoted b and x, then an approximation of the matrix-vector product by a difference

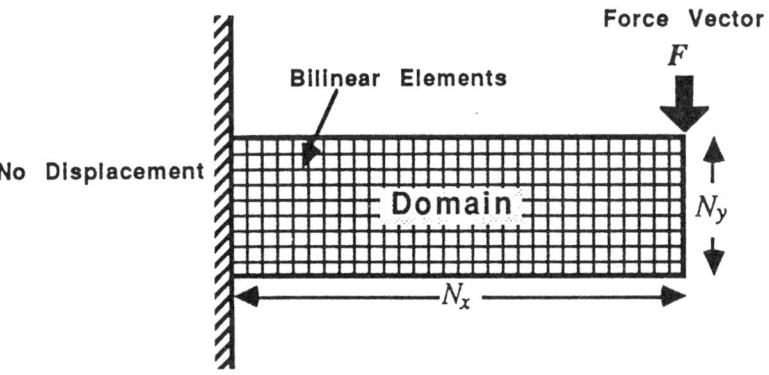

Figure 14.2: Strain Analysis Test Problem.

formula for the directional derivative is

$$Ap_k = \frac{b(x_0 + \epsilon p_k) - b(x_0)}{\epsilon} \tag{14.9}$$

Here, k is the iteration counter and ϵ is the difference parameter. The matrix-free procedure given by (14.9) saves storage of A and repeated calculation of matrix-vector products, at the price of another residual vector calculation at each iteration.

14.4.2 Test Problem

As Figure 14.2 shows, the program computes the deflection of a two-dimensional beam section fixed at one end, subject to a specified point force at the free end. The beam is modeled as a linearly elastic structure with a choice of plane-strain or plane-stress constitutive equations. The quantities N_x and N_y represent the number of finite elements in the x and y directions, respectively.

This application is an established benchmark in structural mechanics, because the resulting matrix equations are poorly-conditioned. High-precision arithmetic and many iterations are needed to reach convergence with standard iterative solution methods.

14.4.3 Parallel Implementation

The parallel implementation of our structural analysis algorithm is summarized by the following host and node processor algorithms (*cf.* Gustafson *et al.* [7]).

ALGORITHM *H*
Host Program for the Strain Analysis Problem

- *H*1 Prompt the user for input values. (Read the dimension of the cube, number of finite elements in the x and y directions, and constitutive model parameters.)

- *H*2 Open a hypercube, and send a copy of the program (Algorithm *N*) to all nodes using a logarithmic fanout.

- *H*3 Send the input values to node 0 (step *N*3).

- *H*4 Create the output header, including input values from step *H*1.

- *H*5 Collect output data from node 0 and print them in the output file. The *message type* shows completion, continuation, or failure of iteration.

- *H*6 If *message type* shows continuation of iterations, repeat step *H*5.

- *H*7 Close the hypercube.   ■

ALGORITHM *N*
Node Program for the Strain Analysis Problem

- *N*1 [Start timer.] Record the time.

- *N*2 [Get node ID.] Execute system call to get this node's process number and the allocated cube dimension.

- *N*3 [Get job parameters.] If the process number is 0, receive job parameters from the host (step *H*3). (This data is then propagated using a logarithmic fanout.)

- *N*4 [Create two-dimensional topology.] Use data from *N*2 and *N*3 to compute processor numbers of nearest neighbors in a two-dimensional subset of the hypercube interconnect, a binary-reflected gray code order (*e.g.*, NCUBE [9]).

$N5$ [Decompose domain.] Based on position in the global domain, compute finite element basis functions, mesh points, first guess x_0, boundary conditions, and mesh-point-to-element mapping.

$N6$ [Start nonlinear iteration.] Begin Newton iteration: calculate residual vector b by 2-by-2-point Gaussian quadrature and save a copy of it as b_0. (We note that on the test problem, which is linear, the code does not return to this step.)

$N7$ [Start linear iteration.] Begin PCG iteration: calculate d, the main diagonal of the iteration matrix A, and d^{-1}. Set projection vector p and vector Ap to zero. Start iteration loop timer.

$N8$ [Start iteration loop.] Set iteration counter i to 1.

$N9$ [Precondition b.] Calculate $z = d^{-1}Ib$, where I is the identity matrix. Exchange boundary values of z. (Send boundary values of z to the nearest neighbor "down," and receive boundary values of z from the nearest neighbor "up" in the gray code decomposition. Send boundary values of z to the nearest neighbor "right," and receive boundary values of z from the nearest neighbor "left" in the gray code decomposition.)

$N10$ [Find directional derivative.] Find "matrix-free" product Az using $q = b(x_0 + \epsilon z)$. Compute q, then exchange and sum boundary values. (Send values of q to the "left" neighbor, receive values of q from the "right" neighbor, and add to the q boundary. Send boundary values of q to the "up" neighbor, and receive boundary values of q from the "down" neighbor, and add to the q boundary.) Compute Az (see (14.9)).

$N11$ [Prepare z for inner product calculation.] Reset "left" and "up" boundary values of z to zero.

$N12$ [Compute inner products.] Calculate local part of $z \cdot b$, $z \cdot Az$, and $p \cdot Az$. Exchange globally and accumulate to get the three global inner products.

$N13$ [Test for convergence.] If $z \cdot b < \delta$, then stop iteration loop timer and proceed with step $N15$. If iteration number i is a multiple of input parameter j and node number is 0, send inner products to host for monitoring progress of calculation.

N14 [Calculate projection length β.] β is $\boldsymbol{z} \cdot \boldsymbol{b}$ divided by $\boldsymbol{z} \cdot \boldsymbol{b}$ from the previous iteration. Then $\boldsymbol{Ap} = \boldsymbol{Az} + \beta\boldsymbol{Ap}$ and $\boldsymbol{p} \cdot \boldsymbol{Ap} = \boldsymbol{z} \cdot \boldsymbol{Az} + 2\beta\boldsymbol{p} \cdot \boldsymbol{Az} + \beta^2\boldsymbol{p} \cdot \boldsymbol{Ap}$.

N15 [Update linear solution.] Calculate α in the PCG algorithm ($\alpha = \boldsymbol{z}{\cdot}\boldsymbol{b}/\boldsymbol{p}{\cdot}\boldsymbol{Ap}$) and do $\boldsymbol{p}$, $\boldsymbol{x}$, and $\boldsymbol{b}$ updates. (That is, $\boldsymbol{p} = \boldsymbol{z}+\beta\boldsymbol{p}$, $\boldsymbol{x} = \boldsymbol{x}+\alpha\boldsymbol{p}$, $\boldsymbol{b} = \boldsymbol{b}-\alpha\boldsymbol{Ap}$.) Increase i by 1 and go to step N9 if i is less than or equal to the maximum number of iterations (specified as a job parameter). If i exceeds that maximum, stop the iteration loop timer, send a message to the host that linear iteration failed and go to step N17.

N16 [Update the nonlinear solution.] If the problem is linear or the Newton iteration has converged, send a message showing this to the host. If the nonlinear iteration count has exceeded the maximum (specified as a job parameter), send a message showing this to the host. Otherwise, calculate a new residual vector $\boldsymbol{b}$ by Gaussian quadrature and go to step N7.

N17 [Complete node timings.] Gather complete and partial (step N6 – N16) timing statistics by a global exchange. Send timing statistics to the host (step H6).   ■

   The parallel PCG and finite element algorithms are based on spatial decomposition. Each processor is assigned a rectangular subdomain within the computational domain of the beam. Each processor subdomain can, in turn, contain thousands of finite elements. During each conjugate gradient iteration, the necessary synchronizations are supplied by the three communication steps, N9, N10, and N12. For P processors, accumulation of inner products (step N12) is done in $\log_2 P$ time by bidirectional exchanges (Saad and Schultz [10]) along successive dimensions of the hypercube, interspersed with summations. The time to complete a global exchange in step N12 for a 10-dimensional hypercube is 7.7 msec, of which 7 msec is message startup (10 reads, 10 writes, 0.35 msec startup per read or write).
   Parallel PCG algorithms have been reported for the CRAY X-MP (Seager [11]) and ELXSI 6400 (Montry and Benner [8]). Another investigation (Barkai *et al.* [2]) showed that PCG can be restructured to reduce both memory/communication traffic and synchronization. We find that by precalculating $\boldsymbol{Az}$ instead of $\boldsymbol{Ap}$ in the iteration loop (step N10), some inner product calculations can be postponed. Hence, all inner products in an iteration can be done with one global exchange. The potential reduction in communication time resulting from fewer global exchanges is 50%. We see

reductions of about 25% for small, communication-dominated problems, *e.g.*, a total of 2048 bilinear elements on 1024 processors. The new algorithm requires the additional storage of a vector of precalculated information.

14.4.4 Communication Cost

The communication cost per iteration (steps $N9$, $N10$, and $N12$) is

$$C_P(n_x, n_y) \;=\; 64(n_x + n_y) + 48 \log_2 P \;, \qquad (14.10)$$

$$M_P(n_x, n_y) \;=\; 8 + 2 \log_2 P \;. \qquad (14.11)$$

These equations do not account for possible overlap of messages. $C_P(n_x, n_y)$ is the number of bytes sent and received per iteration per processor, and $M_P(n_x, n_y)$ is the number of messages sent and received per iteration per processor. P is the number of processors in the ensemble ($P > 1$) and n_x and n_y are the number of gridpoints in the x and y directions local to each processor. The $\log_2 P$ terms result from a global exchange used to calculate global inner products (step $N12$). The other terms arise from nearest neighbor communications (steps $N9$ and $N10$).

14.4.5 Computation Cost

The essential (best serial) operation count in each pass through the iteration loop is

$$\Omega(N_x, N_y) = 111 + 80(N_x + N_y) + 930 N_x N_y \;. \qquad (14.12)$$

Here, N_x and N_y are the number of finite elements in the x and y directions. The actual number of operations done in the parallel version differs from (14.12). To save communication, some operations are done redundantly on all processors, so the parallel operation count is

$$\Omega_P(N_x, N_y) = \; 115P + 5P \log_2 P + 82(N_x + N_y)\sqrt{P} + 930 N_x N_y \;. \quad (14.13)$$

(Note that $\Omega(N_x, N_y) \neq \Omega_1(N_x, N_y)$. Inequality results from exchange and addition of boundary values in step $N10$.) Hence, η_P is a major source of parallel overhead. Table 14.1 shows that the loss of efficiency is as high as 16%.

14.4.6 Measured Performance

Measurements of MFLOPS and job times are given in Tables 14.2 and 14.3 (Gustafson *et al.* [7]). We note that the MFLOPS of Table 14.2 are adjusted

Table 14.1: Operation Efficiency η_P for the Beam Strain Analysis Problem.

Problem size per node	Hypercube Dimension					
	0	2	4	6	8	10
64 by 32	1.0000	0.9978	0.9968	0.9963	0.9961	0.9960
32 by 16	1.0000	0.9956	0.9936	0.9926	0.9921	0.9918
16 by 8	1.0000	0.9910	0.9868	0.9848	0.9838	0.9833
8 by 4	1.0000	0.9808	0.9723	0.9683	0.9663	0.9654
4 by 2	1.0000	0.9578	0.9403	0.9322	0.9284	0.9264
2 by 1	1.0000	0.9040	0.8676	0.8518	0.8444	0.8409

by the operation efficiencies of Table 14.1. That is, only useful MFLOPS are tabulated. Job times for 100 iterations are compared in Table 14.3 to show parallel overhead for each problem size. Time spent outside the iteration loop is also included.

Table 14.2: MFLOPS for the Beam Streain Analysis Problem (64-Bit Arithmetic).

Problem size per node	Hypercube Dimension					
	0	2	4	6	8	10
64 by 32	0.130	0.517	2.07	8.26	33.0	132.0
32 by 16	0.129	0.512	2.04	8.16	32.6	130.0
16 by 8	0.129	0.507	2.01	8.02	32.0	128.0
8 by 4	0.129	0.495	1.94	7.69	30.5	121.0
4 by 2	0.130	0.461	1.75	6.78	26.4	103.0
2 by 1	0.130	0.375	1.30	4.77	17.8	67.1

We note that the number of iterations needed to converge varies as N_x and N_y. All calculations were also run to convergence (norm of residuals reduced by a factor of 1000), except for the job with 2048 by 1024 elements on the full hypercube, which would require over one week to converge.

Execution time drops slightly when the largest problem size is run on four processors instead of serially (Table 14.3, columns 1 and 2). Partitioning the problem among processors gives each processor residual equations involving more similar-sized numbers. This reordering reduces floating point normalization (the shifting of operands so that their binary points align). On the largest problem this effect more than compensates for the greater communication time in going from one to four processors. This result implies that a more efficient ordering of equations for the single processor case is possible (but not necessarily practical).

Table 14.3: Time in Seconds per 100 Iterations for the Beam Strain Analysis Problem.

Problem size per node	Hypercube Dimension					
	0	2	4	6	8	10
64 by 32	1499.8	1499.6	1500.2	1500.8	1501.1	1501.7
32 by 16	378.4	379.3	379.9	380.4	380.9	381.3
16 by 8	95.5	96.4	97.0	97.5	97.9	98.4
8 by 4	24.3	25.1	25.6	26.1	26.5	27.0
4 by 2	6.30	7.07	7.58	8.02	8.45	8.88
2 by 1	1.73†	2.46	2.97	3.40	3.84	4.27

† Result extrapolated from 30 iterations

The scaled speedup result is given by the right graph of Figure 14.3 for problem size fixed at 64 by 32 finite elements per processor. The scaled speedup of 1019 is 99.5% of the ideal. Here, the operation efficiency is 99.6%. Communication overhead, program loading, and startup account for the rest of the efficiency loss (0.1%). When taken to convergence, the accrued overhead (including operation efficiency loss) is about 3000 seconds. The extrapolated uniprocessor execution time for this problem (two million finite elements) is about 20 years. In the parlance of Section 14.2, the serial fraction s' is 0.005.

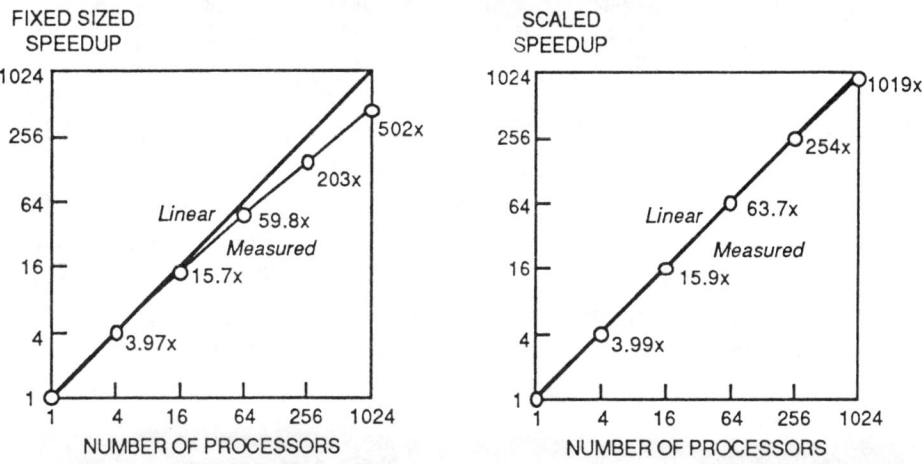

Figure 14.3: Beam Strain Analysis Speedup.

The "constant time" model of speedup implies that global problem size grows as $\approx P^{2/3}$ for this application. For 1024 processors, therefore, we use $1024^{2/3} \approx 100$ times more variables than for the 64 by 32 finite elements on the single processor. This implies 20 by 10 finite elements per processor, or a total of $\approx 400,000$ finite element equations. By interpolating the last column of Tables 14.2 and 14.3, such a problem would run at 128.3 MFLOPS and take 152 seconds per 100 iterations. This performance is 987 times that of a single processor (0.13 MFLOPS). The "constant" run time to convergence for these benchmarks is ≈ 4 hours.

The problem size is constant along the diagonals in Tables 14.1 to 14.3, from top left to lower right. For example, 64 by 32 elements per node on a serial processor is the same as 32 by 16 elements per node on each of four processors. The 64 by 32 grid of finite elements is the largest problem that spans the entire range of hypercube sizes. We note that speedups calculated from the data in Table 14.3 are conservative and higher values of speedup result when calculations, such as those shown in Table 14.4, are run to convergence. The longer execution time amortizes some sources of overhead.

Table 14.4: Beam Strain Analysis Fixed-Sized Problem (2048 Elements).

Hypercube Dimension	Job Time, Seconds	Speed, MFLOPS	Number of Iterations	Fixed-Sized Speedup
0	16278.0	0.128	1087	1.00
2	4097.6	0.508	1088	3.97
4	1038.8	2.00	1087	15.7
6	272.25	7.67	1087	59.8
8	80.01	26.4	1087	203.0
10	32.44	67.1	1088	502.0

The fixed-sized speedup of 502 (Fig. 14.3, left graph) is 49.0% of the ideal linear speedup and implies a serial fractions s of 0.001. Perfect speedup gives an execution time of 15.9 seconds, instead of the measured 32.4 seconds. The difference is the result of three sources of inefficiency. Communication overhead in the parallel version is 10.9 seconds of the total 32.4 seconds reported in Table 14.4. Operation inefficiency is 3.4 seconds, or 15.9% (Table 14.1) of the 21.5 seconds of computation. Program loading and startup take about 2 seconds.

14.5 Summary

We have developed a massively parallel algorithm for a structural analysis problem with a conjugate gradient solver. It appears extensible to higher levels of parallelism than the 1024-processor level validated in this paper. We examined the space of problem size versus ensemble size to assess parallel performance. Scaled and fixed-time models allow one to evaluate ensemble performance over a much broader range of problem sizes than does the fixed-sized model. Efficiency is 0.49 for the fixed-sized model, 0.963 for the fixed-time model, and 0.995 for the scaled model. Operation efficiency, an algorithmic consideration, is the dominant term in accounting for the efficiency loss in scaled speedup.

Even when we are constrained to a fixed-sized problem, the ensemble MFLOPS rate is comparable to the sustained vector MFLOPS rate of typical supercomputers. When mesh sizes are scaled to those typical of the large simulations used in practice, the result is nearly perfect (linear) speedup. These results show the power and future of large processor ensembles in solving problems in structural analysis.

Acknowledgments

We thank Dave Womble of Sandia National Laboratories (SNL) for presenting this paper at the Workshop on Methods and Algorithms for PDE's on Advanced Processors. We also thank John Biffle of SNL for discussions on solid mechanics applications, and Pat Worley of Oak Ridge National Laboratory for supplying a copy of his Ph.D. dissertation, Stanford University (1988), which includes an independent presentation of the fixed-time model.

This work was supported by the Applied Mathematical Sciences Program, U.S. Department of Energy, Office of Energy Research. It was done at Sandia National Laboratories, operated for the U.S. Department of Energy under contract number DE-AC04-76DP00789.

References

[1] Amdahl, G., "Validity of the Single-Processor Approach to Achieving Large-Scale Computer Capabilities," *AFIPS Conference Proceedings*, 30, 483–485, 1967.

[2] Barkai, D., K. J. M. Moriarty, and C. Rebbi, "A Modified Conjugate Gradient Solver for Very Large Systems," *Proceedings of the 1985 International Conference on Parallel Processing*, 284–290, 1985.

[3] Biffle, J. H., "JAC2D-A Two-Dimensional Finite Element Computer Program for the Non-Linear Quasistatic Response of Solids with the Conjugate Gradient Method," *SAND81-0998*, Sandia National Laboratories, Albuquerque, 1984.

[4] Golub, G., and C. Van Loan, *Matrix Computations*, Johns Hopkins Univ. Press, Baltimore, 1983.

[5] Gustafson, J. L., "Author's Response to 'Once Again, Amdahl's Law'," *Communications of the ACM*, 32, 263–264, 1989.

[6] Gustafson, J. L., "Re-evaluating Amdahl's Law," *Communications of the ACM*, 31, 532–533, 1988.

[7] Gustafson, J. L., G. R. Montry, and R. E. Benner, "Development of Parallel Methods for a 1024-Processor Hypercube," *SIAM Journal of Scientific and Statistical Computing*, 9, 609–638, 1988.

[8] Montry, G. R., and R. E. Benner, "Parallel Processing on an ELXSI 6400," *Proceedings of the 2nd International Conference on Supercomputing*, ISI Inc., St. Petersburg, FL, 64–71, 1987.

[9] NCUBE Corporation, *NCUBE Users Manual*, Version P2.1, Beaverton, Oregon, 1987.

[10] Saad, Y., and M. H. Schultz, "Data Communication in Hypercubes," Report YALEU/DCS/RR-428, Yale University, 1985.

[11] Seager, M. K., "Parallelizing Conjugate Gradients for the CRAY X-MP," *Parallel Computing*, 3, 35–47, 1986.

[12] Szabo, G. C., and G. C., Lee, "Derivation of Stiffness Matrices for Problems in Plane Elasticity by Galerkin's Method," *Int. J. Num. Meth. Eng.*, 1, 301-310, 1969.

Chapter 15

Parallel Spectral Element Methods
for Viscous Flow

Paul F. Fischer,[*] *Einar M. Rønquist,*[*]
and Anthony T. Patera[*]

15.1 Introduction

The numerical solution of incompressible fluid flow problems has advanced rapidly in recent years due to simultaneous improvements in algorithms and architectures. In this chapter we describe a parallel spectral element method for solution of the unsteady incompressible Navier-Stokes equations in general three-dimensional geometries. The approach combines high-order spatial discretizations with iterative solution techniques so as to exploit, with high efficiency, currently available medium-grained, distributed-memory parallel computers. We demonstrate the success of this procedure by giving several examples of moderate Reynolds number Navier-Stokes calculations on the Intel vector hypercube.

15.2 Spectral Element Discretizations

We consider here the discretization of the incompressible Navier-Stokes equations in a domain Ω,

---

[*]Massachusetts Institute of Technology, Cambridge, MA.

$$u_t + u \cdot \nabla u \;\; = \;\; \frac{-\nabla p}{\rho} + \nu \nabla^2 u + f \quad \text{in } \Omega, \tag{15.1}$$

$$\nabla \cdot u \;\; = \;\; 0 \quad \text{in } \Omega, \tag{15.2}$$

where $u(x,t)$ is the velocity, $p(x,t)$ is the pressure, x and t are space and time, respectively, $f(x,t)$ is a prescribed force, ρ is the fluid density, and ν is the fluid kinematic viscosity. We assume prescribed initial conditions for the velocity and Dirichlet velocity boundary conditions on the domain boundary $\partial \Omega$.

Our numerical methods for the Navier-Stokes equations (15.1-15.2) are premised upon a 'layered' approach, in which the discretizations and solvers are constructed on the basis of a hierarchy of nested operators proceeding from the highest to the lowest derivatives. For incompressible viscous flow equations the linear self-adjoint elliptic Laplace operator represents the 'kernel' of our Navier-Stokes algorithm insofar that it involves the highest spatial derivatives, and therefore governs the continuity requirements, conditioning and stability of the system. The fully discretized Navier-Stokes equations are typically solved at each time step by performing a series of elliptic solves and preconditioning steps.

To illustrate the essential features of the spectral element discretization [20,26,27], we limit our description to the elliptic 'kernel' represented by the Poisson problem

$$-\nabla^2 u \;\; = \;\; f \quad \text{in } \Omega, \tag{15.3}$$

$$u \;\; = \;\; 0 \quad \text{on } \partial \Omega. \tag{15.4}$$

Our numerical scheme for solving (15.3) and (15.4) is based upon the equivalent variational form: Find $u \in H_0^1(\Omega)$ such that

$$a(u,v) = (f,v) \quad \forall v \in H_0^1(\Omega), \tag{15.5}$$

where

$$\forall \phi, \psi \in L^2(\Omega) \quad (\phi, \psi) \;\; = \;\; \int_\Omega \phi \psi \, d\Omega, \tag{15.6}$$

$$\forall \phi, \psi \in H_0^1(\Omega) \quad a(\phi, \psi) \;\; = \;\; \int_\Omega \nabla \phi \nabla \psi \, d\Omega, \tag{15.7}$$

and the function spaces $L^2(\Omega)$ and $H_0^1(\Omega)$ are defined as

$$L^2(\Omega) \;\; = \;\; \{v \mid \int_\Omega v^2 \, d\Omega < \infty\}, \tag{15.8}$$

$$H_0^1(\Omega) = \{v \mid v \in L^2(\Omega), \nabla v \in L^2(\Omega), v = 0 \ on \ \partial\Omega\}. \quad (15.9)$$

The spectral element method proceeds by first specifying a discretization of the domain Ω as K subdomains,

$$\overline{\Omega} = \cup_{k=1}^{K} \overline{\Omega}^k. \quad (15.10)$$

For reasons of efficiency (see Section 15.3) the subdomains are taken to be quadrilaterals in $I\!\!R^2$ and hexahedra (bricks) in $I\!\!R^3$. For illustrative purposes we consider the simple case where the domain Ω is a two-dimensional region representable by a union of K disjoint squares such that the intersection of two elements is either an edge or a vertex. (A three-dimensional curved-geometry example is given in Section 15.5.) The solution $u(\mathbf{x})$ is approximated using a subspace X_h of $H_0^1(\Omega)$ consisting of all piecewise high-order polynomials of degree $\leq N$,

$$X_h = Y_h \cap H_0^1(\Omega) \quad (15.11)$$

$$Y_h = \{\phi \in L^2(\Omega), \phi \mid_{\Omega^k} \in \mathbf{P}_N(\Omega^k)\}, \quad (15.12)$$

where $\mathbf{P}_N(\Omega^k)$ is the space of all polynomials of degree $\leq N$ in each spatial direction. The spectral element discretization of (15.3-15.4) corresponds to numerical quadrature of the variational form (15.5) restricted to the subspace X_h: Find $u_h \in X_h$ such that

$$a_{h,GL}(u_h, v) = (f, v)_{h,GL} \quad \forall v \in X_h, \quad (15.13)$$

where $(\cdot, \cdot)_{h,GL}$ and $a_{h,GL}(\cdot, \cdot)$ refer to tensor-product Gauss-Lobatto Legendre quadrature of the inner products defined in (15.6-15.7).

It can be shown [20] that the discrete solution u_h converges spectrally fast to the exact solution u for K fixed, $N \to \infty$ ("spectral convergence" [5,14]), with exponential convergence achieved for locally analytic data and solution. This rapid convergence rate derives from the good approximation properties of the polynomial space X_h, and the accuracy associated with the Gauss quadrature and interpolation. The fact that the spectral element method exploits all the regularity of the data and the solution implies that fewer degrees-of-freedom are required to obtain a fixed accuracy compared to a low-order ($N = 1$) finite element method [7,29]. More importantly, if efficient iterative solvers are used (see Section 15.4), the computational cost (operation count) Z required to achieve a discrete solution u_h to a prescribed accuracy ϵ is typically less for a high-order method than a low-order method, as long as ϵ is sufficiently small [27], and the solution sufficiently smooth.

In order to implement (15.13) we require a basis for our high-order polynomial space X_h. The choice of basis does not affect the error estimates. However, it greatly affects the conditioning and the sparsity of the resulting set of algebraic equations, and is critical for the efficiency of parallel iterative solution procedures. We choose an interpolant basis to represent $w_h \in X_h$,

$$w_h(x,y) \mid_{\Omega^k} = \sum_{p=0}^{N} \sum_{q=0}^{N} w_{pq}^k h_p(r) h_q(s)$$

$$x \in \Omega^k \rightarrow (r,s) \in [-1,1]^2, \tag{15.14}$$

where r and s are the local coordinates corresponding to translations of x and y, respectively. The $h_p(z)$ are the one-dimensional N-th order Lagrangian interpolants through the Gauss-Lobatto Legendre points ξ_p ($h_p \in P_N([-1,1]), h_p(\xi_q) = \delta_{pq}$), and w_{pq}^k is the value of w_h at the local node (ξ_p, ξ_q) in element Ω^k. For a function $w_h \in Y_h$ the representation (15.14) is sufficient; however, $w_h \in X_h$ is also required to satisfy the C^0 continuity requirement across elemental boundaries and to satisfy the homogeneous boundary conditions.

The expansion (15.14) is now used in the variational form (15.13), and the test functions are systematically chosen to be unity at one global node and zero at all the other Gauss-Lobatto Legendre points. We then arrive at the discrete matrix statement

$$\sum_{ij}^{\prime k} \sum_{p}^{N} \sum_{q}^{N} (\hat{A}_{ip}^k \hat{B}_{jq}^k + \hat{B}_{ip}^k \hat{A}_{jq}^k) u_{pq}^k = \sum_{ij}^{\prime k} \sum_{p}^{N} \sum_{q}^{N} \hat{B}_{ip}^k \hat{B}_{jq}^k f_{pq}^k$$

$$\forall ij \in \{0,...,N\}^2, \quad \forall k \in \{1,...,K\}, \tag{15.15}$$

where $\hat{A}_{pq}^k$, $\hat{B}_{pq}^k$ are the one-dimensional discrete Laplace operator and mass matrix, respectively.

The two-dimensional direct stiffness operator $\Sigma' : Y_h \rightarrow X_h$ sums contributions from adjacent elements corresponding to local nodes that are physically coincident (enforcing $Y_h \cap H^1$), and masks to zero all local nodes that correspond to boundary points (enforcing $Y_h \cap H_0^1$). The set of algebraic equations (15.15) can also be written as the global matrix statement

$$A u = B f, \tag{15.16}$$

where A is a positive definite symmetric matrix representing the discrete two-dimensional Laplace operator, B is a global diagonal mass matrix, and u and f are global vectors of nodal unknowns and data, respectively.

The preceding approximation of the elliptic problem (15.3-15.4) introduces the key ingredients of the spectral element discretization: variational forms; the piecewise high-degree polynomial approximation space X_h characterized by a discretization pair (K, N); tensor-product spaces, quadratures and bases; and convergence to the exact solution for K fixed and $N \to \infty$. In order to extend this elliptic 'kernel' to solve the unsteady Navier-Stokes equations, two different approaches are pursued.

The first approach uses consistent approximation spaces for the pressure and the velocity in order to ensure a well-posed optimal-order formulation for the saddle Stokes problem [2,12,21,22]. This system is then augmented by a non-dissipative skew-symmetric form of the nonlinear convection operator [27,32] to arrive at the full Navier-Stokes equations. The second approach uses the same (H^1) approximation space for the pressure and the velocity in the context of a splitting scheme (fractional step method) [6,8,17,25,30]. The splitting procedure is attractive as it is both accurate and efficient for sufficiently high Reynolds number flows: it is accurate because the discretization error due to inconsistent pressure boundary conditions decreases as the Reynolds number increases; it is efficient because it involves only the standard discrete H^1 Laplace operator A, rather than the mixed $L^2 - H^1$ pressure operator that results from the use of consistent approximation spaces [18].

15.3 Iterative Solution Procedures

The natural choice of solution algorithms in a parallel environment is an iterative procedure, since such techniques can be both highly local and concurrent. Other considerations such as memory requirements for large three-dimensional problems, variable timestep methods, adaptive mesh refinement, and treatment of time-dependent domains $\Omega(t)$, also suggest the use of iterative solvers. In order to evaluate the efficiency of an iterative approach we must consider two issues: (1) the operation count (or number of clock cycles) per iteration, Z^e; (2) the number of iterations, N_ϵ^A, required to achieve convergence to $\mathcal{O}(\epsilon)$. The total computational cost to "invert" the matrix A in (15.16) is then $Z = N_\epsilon^A Z^e$. In what follows we restrict ourselves to iterative solvers that require only residual evaluation, diagonal preconditioning, and inner-product summation.

Starting with the first issue, the work per iteration is essentially the work to evaluate matrix vector products such as those that appear in (15.15) and (15.16). Due to the tensor-product spaces, tensor-product quadratures, and tensor-product bases described in Section 15.2, these products can be evaluated efficiently using sum-factorization techniques [24] in $\mathcal{O}(KN^{d+1})$ operations in $I\!R^d$. Thus the number of clock cycles required per iteration on a single processor scales as $Z_1^e \sim KN^{d+1}$, an operation count which also applies to general-geometry isoparametric spectral element discretizations of non-separable equations [20].

The second issue is concerned with the convergence rate of an iterative approach. The goal is that the number of iterations N_ϵ^A be independent of the discretization parameter $h = (K, N)$. To invert the symmetric positive definite matrix A in (15.16) we consider two different iterative procedures. The first is a conjugate gradient iteration [13] with $P = diag(A)$ as a diagonal preconditioner. Since this procedure has an order-dependent convergence rate with $N_\epsilon^A \sim K_1 N$ (here K_1 is the number of elements in one spatial direction, [20]), the performance will deteriorate significantly for large problems.

The convergence rate can be improved by the use of an intra-element spectral element multigrid technique [19,27,28,33]. The key ingredients are: variational forms allowing for consistent derivation of restriction and prolongation operators; nested spaces defined by varying the polynomial degree, N, within K fixed elements; and highly parallelizable P- preconditioned Jacobi iteration as a smoother. This procedure results in an order-independent convergence rate in $I\!R^1$ ($N_\epsilon^A \sim \mathcal{O}(1)$), and only a weak N-dependence in $I\!R^2$ and $I\!R^3$ ($N_\epsilon^A \sim N^{1/2}$).

The iterative solution procedures for the unsteady Navier-Stokes equations rely strongly on an efficient solver for the elliptic kernel. However, the particular choice of algorithm depends on the discretization approach. If consistent approximation spaces are used for the velocity and the pressure (see Section 15.2), the implicit Stokes operator is best inverted by a nested Uzawa conjugate gradient/multigrid decoupling scheme [3,4,18], with the nonsymmetric convective operator treated explicitly [27]. The splitting approach is also a semi-implicit scheme (explicit convection/implicit Stokes), requiring only $d + 1$ standard (H^1) elliptic solves in $I\!R^d$ each timestep (one for the pressure, and one for each velocity component) [8,17,25].

15.4 Parallelism

The spectral element discretizations, bases, and iterative solvers of the previous sections are constructed so as to admit a native, geometry-based parallelism, in which each spectral element (or group of spectral elements) is mapped to a separate *processor/memory* unit, with the individual processor/memory units being linked by a relatively sparse *communication* network. This distributed-memory conceptual architecture is naturally suited to the spectral element discretization in that it provides a tight, structured coupling within the dense elemental constructs (vectorizable), while simultaneously maintaining generality and concurrency at the level of the unstructured macro-elemental skeleton (parallelizable). We now review some of the key issues related to parallel spectral element solution of the unsteady Navier-Stokes equations [26]. In particular we consider the iterative solution of the $h = (K, N)$ elliptic spectral system (15.16) on M independent parallel processor/memory units.

Two new aspects of our parallel methods are the use of high-order discretizations to reduce communication through reduced degrees-of-freedom, and the construction of algorithms that do not require *global* structure to be effective. Iterative solution of (15.16) requires repeated evaluation of global matrix-vector products of the form $A\,u$. Yet it is clear from (15.15) that this matrix-vector product can be evaluated in an essentially local manner, since the inner most summations range over gridpoints within elements. The parallel implementation, therefore, reduces to distributing the K elemental data and matrix coefficients to M independent processors, which in turn evaluate the local, element based residuals simultaneously. (See Figures 15.1(a) and 15.1(b).) This (incomplete) residual evaluation is computationally the most expensive part of one single iteration (measured in processor clock cycles), and it can be shown that the leading term will be $Z_M^e \sim KN^{d+1}/M$ in $\mathbf{R}^d$, or $Z_M^e \sim Z_1^e/M$, where Z_1^e is the work on a single processor, and K is assumed to be a multiple of M.

In addition to the computational cost we must also consider the communication overhead. First, the incomplete residual evaluation is completed by summing nodal values corresponding to shared degrees-of-freedom at element interfaces, as denoted by the direct stiffness summation operator, Σ'. In Figures 15.1(a) through 15.1(f) we show diagramatically how the direct stiffness summation is effected on a distributed processor by exchanging element edge data between processor pairs corresponding to adjacent element pairs. Once exchanged, the buffered edge data can be added to the associated residual values of the resident elements. Neighboring elements couple only on elemental boundaries, which implies that only $(\mathcal{O}(N^{d-1}))$ surface data need to be communicated per element/processor in order to ensure C^0-continuity of the discrete solution.

For many problem topologies, an ordered exchange sequence can be employed to ensure that the correct residual will be computed at element vertices (corners) without vertex-specific data exchanges, resulting in an efficient parallel direct stiffness summation. Not all topologies admit such economy, however, and it is crucial that generality not be sacrificed for reasons of efficiency. In such cases, vector reduction is used in a gather-scatter fashion to sum the few vertex values that are not correctly updated via the ordered ex-

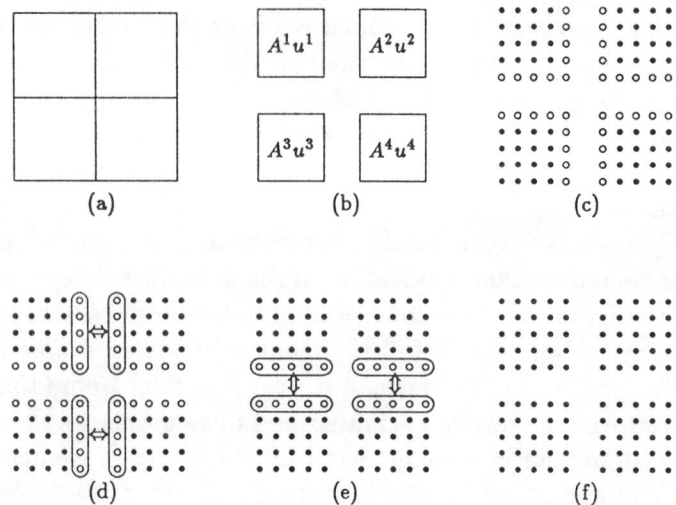

Figure 15.1: Computation of residual vector for regular geometry in $I\!\!R^2$: (a) four element mesh; (b) simultaneous (parallel) computation of incomplete residual $\tilde{r}^k = A^k u^k$; (c) nodal content of $\tilde{r}^k$ is denoted by circles — solid circles indicate correct residuals, open circles indicate values requiring contributions from neighboring elements; (d) and (e) bi-directional exchange and sum sequence, ψ, which affects the completed residual at all nodes including the corner nodes; (f) completed residual, $r^k = \sum' A^k u^k$.

change procedure. Although the vector reduction invokes a global construct, the amount of data involved is small ($\mathcal{O}(N^{d-2})$). Thus the incurred cost is tolerable.

For conjugate gradient solution of (15.16), the inner-product summation also requires inter-processor communication, as data that are distributed across an ensemble of processors need to be reduced to a single value. Such global vector reductions are efficiently computed on distributed-memory architectures by first evaluating M *local* inner-products, and then summing the partial terms via a binary spanning tree in $\mathcal{O}(logM)$ communication cycles [29]. Because of their global nature, the execution time required for inner-product calculations does not M-parallelize, and is directly dependent upon system size. Consequently, they can become the dominant part of the calculation as the number of processors grows large. The balance between the computation time (Z_M^e) and the communication time ($logM$) determines the optimal granularity for the problem.

For illustration, we plot in Figure 15.2 a measure of the solution time (wall clock time in our single user environment) versus number of processors for some fixed problem. The τ_{comp} curve, corresponding to the total computing effort (operations) required to solve the problem, decreases as the number of processors assigned to the task increases, assuming the task is M-partitionable. The parallel overhead costs are illustrated by the τ_{ds} and τ_{ip} curves, which represent the direct stiffness summation and inner-product communication times, respectively. The total solution time, τ_{sol}, is the sum of the three cost terms: τ_{comp}, τ_{ds}, and τ_{ip}. Note that, in general, it is not possible to span the entire range of processors, $M = 1, 2, \ldots, M_{max}$, for a single problem (K, N, d) due to the coarse grained nature of the spectral element discretization. The lower bound is controlled by the ratio KN^d/M, which is limited by the memory on each processor. The upper bound is $M \leq K$, since elements are assumed to be indivisible entities.

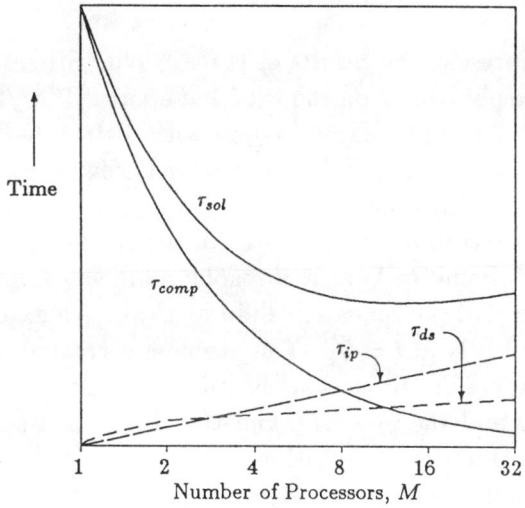

Figure 15.2: Illustration of computing costs associated with the parallel conjugate gradient algorithm, measured in wall clock time. The growth of the inner-product time illustrates the limiting nature of *global* operations, resulting in a minimum solution time for a number of processors $M_0 < \infty$.

Lastly, we make some comments regarding the parallel efficiency, $\eta \equiv \tau_{comp}/\tau_{sol}$. It should be noted that using η as a performance measure can be misleading, since any algorithm or architecture improvements that directly decrease τ_{comp} will reduce τ_{sol}, but will also yield reduced efficiency. It is, nonetheless, of interest to analyze the efficiency for a fixed algorithm-architecture coupling and for a fixed ratio, K/M. Such an analysis addresses the question of what performance can be obtained if the problem size is doubled and the number of processors is doubled. It can be shown that there is only a small logarithmic degradation in the efficiency as the problem size and machine size grow simultaneously. Such performance has been observed empirically in our experience with the iPSC/2-VX. A typical three-dimensional flow calculation with $N = 10$ and $K/M = 2$ runs at roughly 75 percent efficiency and 10 to 11 MFLOPS on four processors. Similar performance is obtained for larger systems in which $K/M = 2$; for example, 16 processors yield 44 MFLOPS, 32 processors yield 80 MFLOPS, and 64 processors yield 160 MFLOPS.

For more details regarding the parallel implementation see [26]. Related work on parallel partial differential equation solution by domain decomposition is described in [27-30].

15.5 Navier-Stokes Calculations

In this section we present the results of three Navier-Stokes calculations. All the calculations are performed on the Intel hypercube iPSC/2-VX parallel computer. This machine is a 386-based system with distributed memory and pipelined communication routing, and node vector hardware capable of peak performance of 10 MFLOPS/node.

The first example we consider is the classical problem of external startup flow past a cylinder of diameter D at a Reynolds number of $R = U_\infty D/\nu = 100$. The cylinder is initially in quiescent fluid at $t = 0$. An external uniform flow U_∞ is imposed abruptly at $t = 0_+$. The problem is treated with the spectral element Navier-Stokes discretizations based on consistent approximation spaces for the velocity and the pressure, and the Uzawa conjugate gradient-based iterative solvers (multigrid algorithms are not yet implemented on the hypercube). Figure 15.3(a) shows a good comparison between the numerical prediction of the recirculation zone length at early times, and the experimental observations of [15]. At later times the familiar unsteady von Karman vortex street forms as shown in Figure 15.3(b).

The second problem is again the cylinder startup problem, but now at a Reynolds number $R = 1000$. This calculation is based on the splitting scheme with conjugate gradient solution of the elliptic subproblems. Figure 15.4(a) shows the spectral element discretization of the computational domain into $K = 52$ elements, together with instantaneous streamlines at an early non-dimensional time of $U_\infty t/D = 3$. Figure 15.4(b) shows a more detailed view of the vortex structure that forms behind the cylinder. This result, and in particular, the location of the small secondary vortex, is in excellent agreement with the experimental results provided by [31].

The last example is a three-dimensional flow past a cylinder mounted on a flat plate. In the flow (x) direction the velocity profile $u = U_*[1 - 16(1/2 - z/H)^4], z \leq H/2$, is imposed at the inflow plane, while outflow boundary conditions are imposed at the exit plane. Periodic boundary conditions are imposed in the y-direction, no-slip wall boundary conditions are imposed on the plate ($z = 0$), and symmetry boundary conditions are assumed on the top plane ($z = H/2$). The Reynolds number based on the symmetry plane inflow velocity (U_*) and cylinder diameter, D, is $R = 1000$; the geometric ratio D/H is unity for this calculation. Figure 15.5a shows the three-dimensional horseshoe vortex structure that forms around the cylinder at a non-dimensional time $U_*t/D = 4$, while Figure 15.5b shows the more detailed flow structure close to the cylinder at a later time $U_*t/D = 9$. The computational results are in good qualitative agreement with the experimental results of [1].

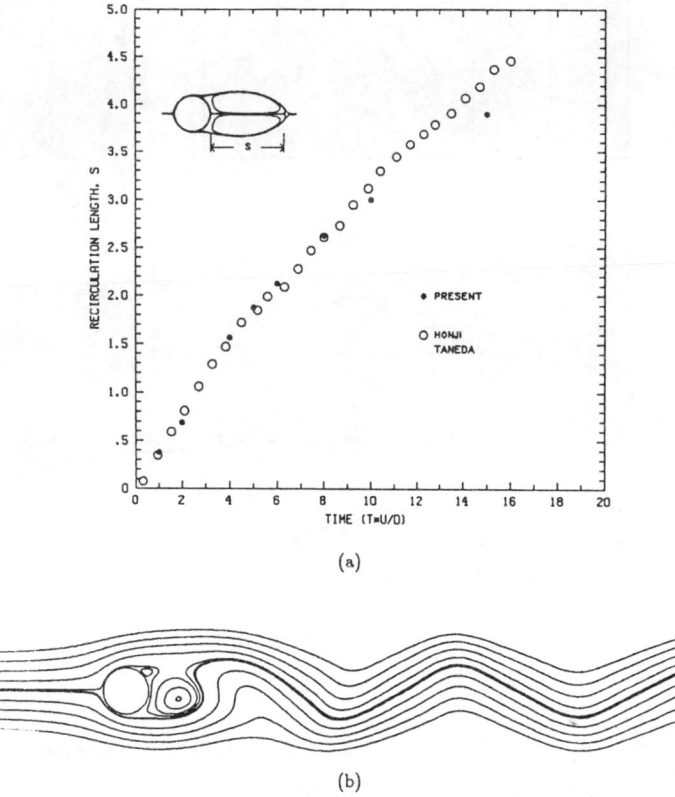

(a)

(b)

Figure 15.3: Startup flow past a cylinder at a Reynolds number $R = U_\infty D/\nu = 100$. (a) Comparison of early time nondimensional recirculation zone length with experimental measurements of [15]. (b) Instantaneous streamlines showing the von Karman vortex street at a nondimensional time of 110.

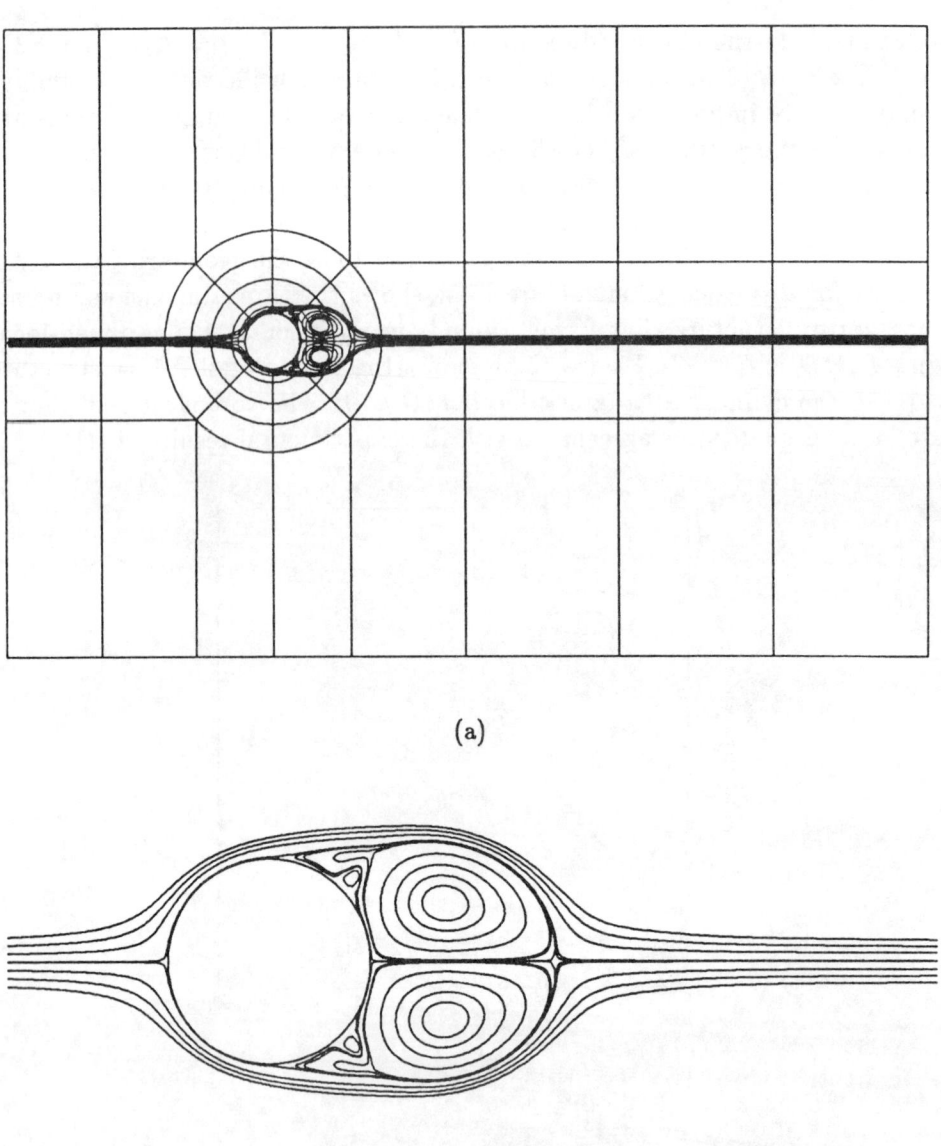

(a)

(b)

Figure 15.4: Startup flow past a cylinder at a Reynolds number $R = U_\infty D/\nu$ = 1000. (a) Spectral element mesh ($K = 68$, $N = 10$) and instantaneous streamlines at a nondimensional time of $U_\infty t/D = 3$. (b) Enlarged view reveals small secondary vortices which are are also observed experimentally [31].

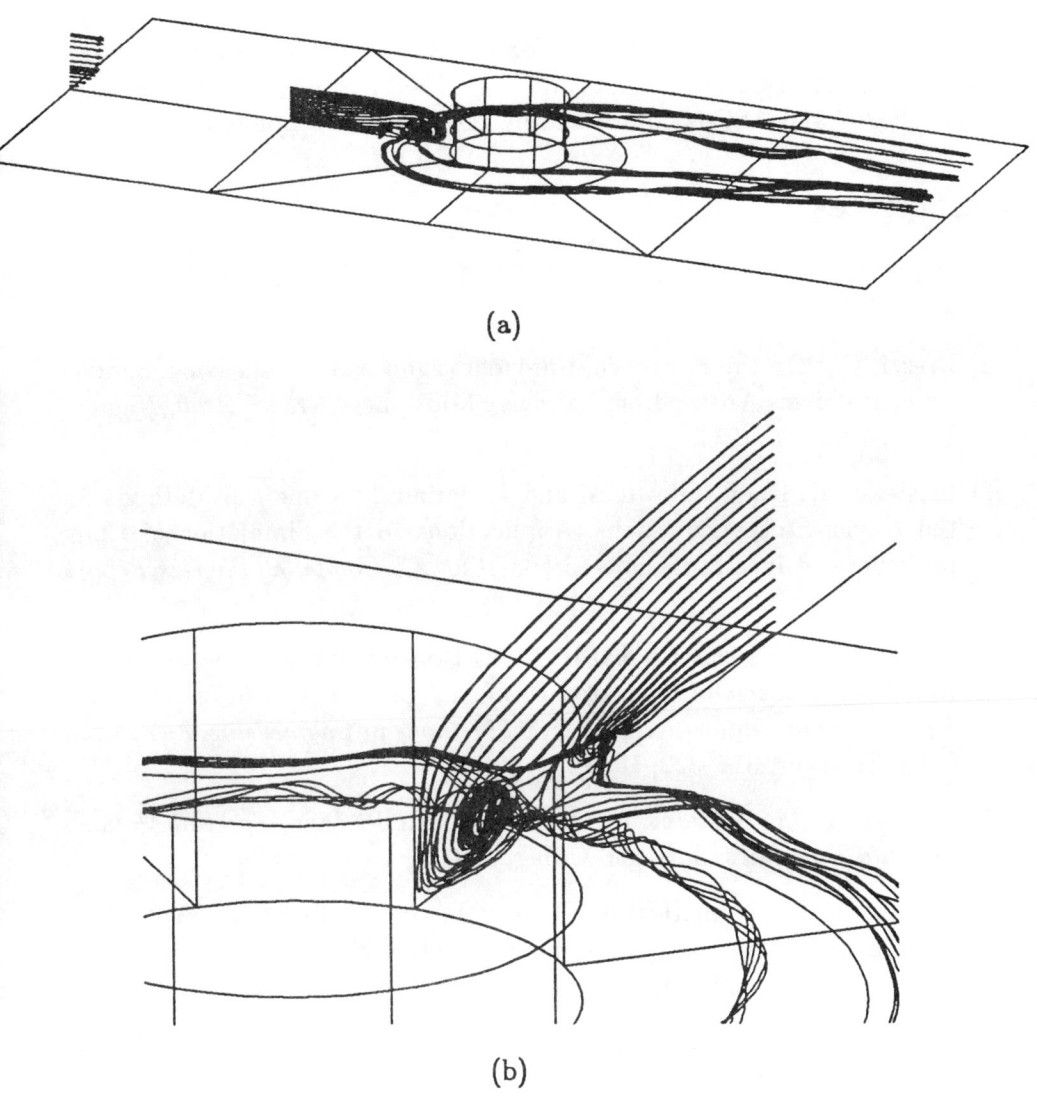

(a)

(b)

Figure 15.5: Three-dimensional flow past a cylinder mounted on a flatplate at a Reynolds number $R = 1000$ based on maximum inflow velocity, U_*, and cylinder diameter, D. (a) Mesh, inflow velocityprofile, and three-dimensional horseshoe vortex structures shown at an early time $U_* t/D = 4$. (b) Detailed flow structure close to the cylinder reveals the presence of two horseshoe vortices at time $U_* t/D = 9$, in good qualitative agreement with the experiments of [1].

Acknowledgments

This work was supported by the ONR and DARPA under contracts N00014-85-K-0208, N00014-87-K-0439, and N00014-88-K-0188, by NSF under Grants DMC-8704357 and ASC-8806925, and by Intel Scientific Computers.

References

[1] Baker, C. J., "The Laminar Horseshoe Vortex," *J. Fluid Mech.*, 95, 347–368, 1979.

[2] Brezzi, F., "On the Existence, Uniqueness and Approximation of Saddle-Point Problems Arising from Lagrange Multipliers," *Rairo Anal. Numer.*, 8, R2, 129–151, 1974.

[3] Bristeau, M. O., R. Glowinski and J. Periaux, "Numerical Methods for the Navier-Stokes Equations. Applications to the Simulation of Compressible and Incompressible Viscous Flows," *Computer Physics Report*, to appear.

[4] Cahouet, C., and J. P. Chabard, "Multi-Domain and Multi-Solvers Finite Element Approach for the Stokes Problem." in *Proc. Fourth International Symposium on Innovative Numerical Methods in Engineering*, R. P. Shaw (ed.), Springer, 317–322, 1986.

[5] Canuto, C., M. Hussaini, A. Quarteroni and T. Zang, *Spectral Methods in Fluid Dynamics*, Springer-Verlag, 1987.

[6] Chorin, A. J., "Numerical Solution of Incompressible Flow Problems," in *Studies in Numerical Analysis*, J. M. Ortega and W. C. Rheinboldt (eds.), SIAM, 64–71, 1970.

[7] Ciarlet, P., *The Finite Element Method for Elliptic Problems*, North-Holland, Amsterdam, 1978.

[8] Fischer, P. F., *Parallel Spectral Element Methods for the Incompressible Navier-Stokes Equations*, Ph.D. Thesis, Massachusetts Institute of Technology, in progress, 1989.

[9] Fischer, P. F., and A. T. Patera, "Parallel Spectral Element Solution of the Stokes Problem," *J. Comp. Phys.*, submitted, 1989.

[10] Fox, G. C. and S. W. Otto, "Concurrent Computation and the Theory of Complex Systems," in *Hypercube Multiprocessors* (Edited by M.T. Heath), SIAM, Philadelphia, 244–268, 1986.

[11] Glowinski, R., G. Golub, G. Meurant and J. Periaux (eds), *Proceedings of the First International Conference on Domain Decomposition Methods for Partial Differential Equations*, SIAM, Philadelphia, 1987.

[12] Girault, V., and P. A. Raviart, *Finite Element Approximation of the Navier-Stokes Equations*, Springer, 1986.

[13] Golub, G. H., and C. F. Van Loan, *Matrix Computations*, Johns Hopkins University Press, Baltimore, Maryland, 1983.

[14] Gottlieb, D., and S. A. Orszag, *Numerical Analysis of Spectral Methods*, SIAM, Philadelphia, 1977.

[15] Honji, H., and A. Taneda, "Unsteady Flow Past a Circular Cylinder," *J. Phys. Soc. Japan*, 27, 6, 1668–1677, 1969.

[16] Keyes, D. E., and W. D., Gropp, in *Proceedings of the Second International Conference on Domain Decomposition Methods for Partial Differential Equations* T. Chan (ed.), SIAM, Philadelphia, to appear.

[17] Korczak, K. Z., and A. T., Patera, "An Isoparametric Spectral Element Method for Solution of the Navier-Stokes Equations in Complex Geometry," *J. Comput. Phys.*, 62, 361–382, 1986.

[18] Maday, Y., D. Meiron, A. T. Patera and E. M. Rønquist, "Analysis of Iterative Methods for the Steady and Unsteady Stokes Problem: Application to Spectral Element Discretizations," *J. Comp. Phys*, submitted, 1989.

[19] Maday, Y., and M. Muñoz, "Spectral Element Multigrid, II. Theoretical Justification," *J. Sci. Comput.*, to appear, 1989.

[20] Maday, Y., and A. T. Patera, "Spectral Element Methods for the Navier-Stokes Equations," in *State of the Art Surveys in Computational Mechanics* A. K. Noor (ed.), ASME, Winter Annual Meeting, Anaheim, CA, 1986.

[21] Maday, Y., A. T. Patera and E. M. Rønquist, "A Well-Posed Optimal Spectral Element Approximation for the Stokes Problem," *SIAM J. Numer. Anal.*, to appear.

[22] Maday, Y., A. T. Patera and E. M. Rønquist, "Optimal Legendre Spectral Element Methods for the Multi-Dimensional Stokes Problem," in preparation.

[23] McBryan, O. A., and E. F. van de Velde, in *Selected Papers from the Second Conference on Parallel Processing for Scientific Computing* C. W. Gear and R. G. Voigt (eds.), SIAM, Philadelphia, S227–S287, 1987.

[24] Orszag, S. A., "Spectral Methods for Problems in Complex Geometries," *J. Comput. Phys.*, 37, 70–92, 1980.

[25] Orszag, S. A., M. Israeli and M. O. Deville, "Boundary Conditions for Incompressible Flows," *J. Sci. Comput.*, 1, 75–111, 1986.

[26] Patera, A. T., "A Spectral Element Method for Fluid Dynamics; Laminar Flow in a Channel Expansion," *J. Comp. Phys.*, 54, 468–488, 1984.

[27] Rønquist, E.M., *Optimal Spectral Element Methods for the Unsteady Three-dimensional Incompressible Navier-Stokes Equations*, Ph.D. Thesis, Massachusetts Institute of Technology, 1988.

[28] Rønquist, E.M., and A. T. Patera, "Spectral Element Multigrid: I. Formulation and Numerical Results," *J. Sci. Comp.*, 2, 4, 389–406, 1987.

[29] Strang, G. and G. Fix, *An Analysis of the Finite Element Method*, Prentice-Hall, 1973.

[30] Temam, R., *Navier-Stokes Equations. Theory and Numerical Analysis*, North-Holland, Amsterdam, 1984.

[31] Van Dyke, M., *An Album of Fluid Motion*, The Parabolic Press, Stanford, California, 36, 1982.

[32] Zang, T. A., "On the Rotation and Skew-Symmetric Forms for Incompressible Flow Simulations," *Applied Num. Math.*, to appear.

[33] Zang, T. A., Y. S. Wong and M. Y. Hussaini, "Spectral Multigrid Methods for Elliptic Equations II," *J. Comp. Phys*, 54, 489–507, 1982.

Chapter 16

A Multiplier/Element by Element Method for a Class of Nonlinear Boundary Value Problems

*R. Glowinski**

16.1 Introduction

The main goal of this chapter is to describe the application of a decomposition principle involving multipliers to a class of nonlinear boundary value problems. The finite dimensional implementation of this decomposition principle — via finite elements for example — leads quite naturally to a two-step algorithm. One of the steps consists of the elementwise solution of problems of very small dimensions. This basic property may be essential to successful implementation of the related algorithms on parallel and/or vector machines. Two examples will be discussed in some detail. The first is associated with a visco-plastic flow (Sections 16.2, 16.7 and 16.9), and the other one with Yang Mills equations (the Skyrme model; cf Section 16.10).

16.2 Formulation of the Problems

Let Ω be a bounded domain of $I\!R^N$, and denote by Γ its boundary. We consider the following class of nonlinear, second order boundary value problems (for simplicity, we consider Dirichlet boundary conditions only; also, we

---

*University of Houston, Houston, TX.

shall not discuss the existence theory for solutions):

$$- \nabla \cdot (A(x)\nabla u) + f(x, u) - \nabla \cdot B(x, \nabla u) = 0 \text{ in } \Omega, \qquad (16.1)$$

$$u = g \text{ on } \Gamma. \qquad (16.2)$$

In (16.1) and (16.2),

1. $A(x)$ is *a.e.* on Ω an $N \times N$ matrix, possibly non symmetric, but uniformly positive definite over Ω; that is, $\exists \, \alpha > 0$ such that

$$A(x)\xi \cdot \xi \geq \alpha |\xi|^2, \quad \text{a.e. } x \in \Omega, \forall \, \xi \in I\!R^N,$$

2. f is a (possibly non-smooth) function of x, and u, taking its values in $I\!R$ (or $\bar{I\!R}$),

3. B is a (possibly non-smooth) function of $\{x, q\} \in I\!R \times I\!R^N$ taking its values in $I\!R^N$ (or $\bar{I\!R}^N$).

Here the following notation has been used:

$$a \cdot b = \sum_{i=1}^{N} a_i b_i \; \forall \, a = \{a_i\}_{i=1}^{N}, \; b = \{b_i\}_{i=1}^{N} \in I\!R^N; \; |a| = (a, a)^{1/2},$$

$$\bar{I\!R} = I\!R \cup \{\pm\infty\}, \nabla = \left\{\frac{\partial}{\partial x_i}\right\}_{i=1}^{N}.$$

We are particularly concerned with those situations where Newton's and conjugate gradient methods may perform very poorly due to the nonsmoothness of f and/or B. An example of (16.1) is given by the following visco-plastic flow problem.

Example 1:   (Flow of a visco-plastic material in a cylindrical pipe)

$$- \mu \nabla^2 u - g \nabla \cdot \left(\frac{\nabla u}{|\nabla u|}\right) = f \text{ in } \Omega, \qquad (16.3)$$

$$u = 0 \text{ on } \Gamma. \qquad (16.4)$$

In (16.3) and (16.4), $\Omega(\subset I\!R^2)$ is the cross-section of the pipe, u is the axial velocity of the flow, $\mu(> 0)$ is the viscosity parameter, g is the plasticity yield, and f (usually a constant) is the linear decay of pressure in the axial direction. This problem is discussed in detail in, e.g., [5,7,8,9,10]. In this visco-plastic flow B is not well behaved and will lead to difficulties in the approximate solution. It is clear, however, that the problem can be approximated by a smoother one such as

$$-\mu\nabla^2 u_\epsilon - g\nabla \cdot \left(\frac{\nabla u_\epsilon}{\sqrt{\epsilon^2 + |\nabla u_\epsilon|^2}}\right) = f \text{ in } \Omega, \qquad (16.5)$$

$$u_\epsilon = 0 \text{ on } \Gamma. \qquad (16.6)$$

where $\epsilon > 0$ is a (small) parameter. Suppose that $f \in L^2(\Omega)$. We can then prove that $\|u_\epsilon - u\|_{H^1(\Omega)} \leq C\sqrt{\epsilon}$, which justifies the above approximation (in fact if $f = \text{const.}$ and Ω is a disk we can prove that $\|u_\epsilon - u\|_{H^1(\Omega)} \leq C_\epsilon |\log \epsilon|^{1/2}$).

Actually, the above "cure" is not at all miraculous because a good approximation requires a small ϵ, but in that case the regularized problem (16.5), (16.6) is poorly conditioned.

Indeed with the methods to be described in the following sections of this chapter, we shall be able to solve problem (16.3), (16.4) (and also (16.5), (16.6)) at a small cost (typically 3 to 5 times the cost of solving a linear Dirichlet problem, like the one obtained when $g = 0$).

16.3 Variational Formulation

Let us define the following spaces

$$H^1(\Omega) = \left\{v | v \in L^2(\Omega), \frac{\partial v}{\partial x_i} \in L^2(\Omega), \forall i = 1, 2, \ldots, N\right\}, \qquad (16.7)$$

$$H_0^1(\Omega) = \left\{v | v \in H^1(\Omega), v = 0 \text{ on } \Gamma\right\}, \qquad (16.8)$$

$$H_g^1(\Omega) = \left\{v | v \in H^1(\Omega), v = g \text{ on } \Gamma\right\}. \qquad (16.9)$$

In (16.7), $\frac{\partial v}{\partial x_i}$ is interpreted in the sense of distributions; however, if g is sufficiently smooth then $H_g^1(\Omega)$ is not empty. Taking $v \in H_0^1(\Omega)$, multiplying both sides of (16.1) by v and integrating by parts we obtain the following variational formulation

$$\int_\Omega A(x)\nabla u \cdot \nabla v dx + \int_\Omega f(x, u)v dx + \int_\Omega B(x, \nabla u) \cdot \nabla v dx = 0,$$

$$\forall v \in H_0^1(\Omega), u \in H_g^1(\Omega). \qquad (16.10)$$

Problem (16.10) is equivalent to problem (16.1), but is better suited to the subsequent treatment.

16.4 A Decomposition Principle

Let us define now W_0 and W_g by

$$W_0 = \left\{ \{v, q\} \,|\, v \in H_0^1(\Omega),\ q \in \left(L^2(\Omega)\right)^N,\ \nabla v - q = 0 \right\}, \qquad (16.11)$$

$$W_g = \left\{ \{v, q\} \,|\, v \in H_0^1(\Omega),\ q \in \left(L^2(\Omega)\right)^N,\ \nabla v - q = 0 \right\}, \qquad (16.12)$$

respectively. Problems (16.1) and (16.10) are then clearly equivalent to

$$\int_\Omega A(x) \nabla u \cdot \nabla v\, dx + \int_\Omega f(x, u) v\, dx + \int_\Omega B(x, p) \cdot q\, dx = 0,$$

$$\forall \; \{v, q\} \in W_0;\ \{u, p\} \in W_g; \qquad (16.13)$$

A possible variant of (16.13) is

$$\int_\Omega A(x) p \cdot q\, dx + \int_\Omega f(x, u) v\, dx + \int_\Omega B(x, p) \cdot q\, dx = 0,$$

$$\forall \; \{v, q\} \in W_0;\ \{u, p\} \in W_g. \qquad (16.14)$$

Before showing how to exploit formulation (16.13) (or (16.14)) let us consider the following finite dimensional problem: Find x satisfying $Bx = c$ and

$$F(x) \cdot y = 0, \quad \forall\, y \in \text{Ker } B, \qquad (16.15)$$

where in (16.15), F is some function from $I\!R^N$ to $I\!R^N$, B is an $M \times N$ matrix and $c \in I\!R^M$. Assuming that problem (16.15) has at least one solution, we may easily prove that solving (16.15) is equivalent to finding a pair $\{x, \lambda\} \in I\!R^N \times I\!R^M$ satisfying

$$F(x) + B^T \lambda \;=\; 0,$$

$$Bx \;=\; c \qquad (16.16)$$

where λ is a (Lagrange) multiplier. Similarily, (16.13) implies, at least formally, the existence of $\lambda \in (L^2(\Omega))^N$ such that

$$\int_\Omega A \nabla u \cdot \nabla v\, dx + \int_\Omega f(x, u) v\, dx + \int_\Omega B(x, p) \cdot q\, dx$$

$$+ \int_\Omega \lambda \cdot (\nabla v - q)\, dx = 0, \quad \forall\, v \in H_0^1(\Omega),\ \forall\, q \in (L^2(\Omega))^N,$$

$$\nabla u - p = 0;\ u \in H_g^1(\Omega),\ p,\ \lambda \in (L^2(\Omega))^N. \qquad (16.17)$$

The corresponding variant associated with (16.14) is obvious.

From a practical point of view, we observe that (16.17) is equivalent to

$$\int_\Omega A\nabla u \cdot \nabla v dx + \int_\Omega f(x,u)v dx + \int_\Omega B(x,p)\cdot q dx$$

$$+ r\int_\Omega (\nabla u - p)\cdot(\nabla v - q)dx + \int_\Omega \lambda \cdot (\nabla v - q)dx = 0,$$

$$\forall v \in H_0^1(\Omega), \quad \forall q \in (L^2(\Omega))^N,$$

$$\nabla u - p = 0; \ u \in H_g^1(\Omega), \ p, \ \lambda \in (L^2(\Omega))^N, \tag{16.18}$$

where, in (16.18), r is a positive parameter. (If a negative parameter r is better suited to the problem the modifications are straight-forward.)

Problem (16.18) is equivalent to the following system: Find $u \in H_g^1(\Omega)$ and $p \in (L^2(\Omega))^N$ such that

$$\int_\Omega A\nabla u \cdot \nabla v dx + r\int_\Omega \nabla u \cdot \nabla v dx + \int_\Omega f(x,u)v dx$$

$$= r\int_\Omega p \cdot \nabla v dx - \int_\Omega \lambda \cdot \nabla v dx, \quad \forall v \in H_0^1(\Omega), \tag{16.19}$$

$$\int_\Omega B(x,p)\cdot q dx + r\int_\Omega p \cdot q dx$$

$$= r\int_\Omega \nabla u \cdot q dx + \int_\Omega \lambda \cdot q dx, \quad \forall q \in (L^2(\Omega))^N, \tag{16.20}$$

$$\nabla u - p = 0 \tag{16.21}$$

In "non variational form", system (16.19) - (16.21) can also be written

$$-\nabla \cdot (A\nabla u) - r\nabla^2 u + f(x,u) = \nabla \cdot \lambda - r\nabla \cdot p \text{ in } \Omega, \tag{16.22}$$

$$B(x,p) + rp = \lambda + r\nabla u \quad a.e. \text{ on } \Omega, \tag{16.23}$$

$$\nabla u - p = 0. \tag{16.24}$$

with $u = g$ on Γ.

16.5 Algorithms

There are indeed quite a few algorithms exploiting the above Decomposition Principle, as shown in, e.g., [7,8,9]; a simple one is given by:

$$\lambda^0 \in \left(L^2(\Omega)\right)^N, \ u^{-1} \in H_g^1(\Omega) \text{ given}; \tag{16.25}$$

for $n \geq 0$, assuming that λ^n and u^{n-1} are known, compute p^n, u^n as the solution of

$$B(x, p^n(x)) + r p^n(x) = r \nabla u^{n-1} + \lambda^n(x), \quad a.e. \text{ on } \Omega, \qquad (16.26)$$

$$-\nabla \cdot (A \nabla u^n) - r \nabla^2 u^n + f(x, u^n) = \nabla \cdot \lambda^n - r \nabla \cdot p^n \text{ on } \Omega,$$

$$u^n = g \text{ on } \Gamma, \qquad (16.27)$$

and then update λ^{n+1} by

$$\lambda^{n+1} = \lambda^n + r(\nabla u^n - p^n), \qquad (16.28)$$

The above algorithm has many variants, some of them discussed in [3,7,8,9]; an obvious one is to exchange the roles of u^n and p^n, solving first the u^n problem, and then the p^n problem. Actually, we can increase the robustness of algorithm (16.25)—(16.28) by doing several iterations of (16.26), (16.27) before updating the multiplier λ^n. The convergence of the above algorithm, and of some of its variants, is discussed in [7,8,9].

16.6   Time Relaxation Interpretation

The following analysis of algorithms (16.25)—(16.28) is quite formal. It can however be rigorously justified if for x given, $v \to f(x, v)$ and $q \to B(x, q)$ are monotone operators (possibly multivalued); see [7,9,12].

Take $r = 0$ in (16.24). We have then

$$-\nabla \cdot (A \nabla u) + f(x, u) = \nabla \cdot \lambda \text{ in } \Omega, \quad u = g \text{ on } \Gamma, \qquad (16.29)$$

$$B(x, p) = \lambda(x) \text{ a.e. on } \Omega, \qquad (16.30)$$

$$\nabla u - p = 0. \qquad (16.31)$$

Let $\mathcal{A}$ be the operator defined over H_g^1 by

$$\mathcal{A}(v)(x) = -\nabla \cdot (A \nabla v) + f(x, v). \qquad (16.32)$$

Similarly, define $\mathcal{B}$ by

$$\mathcal{B}(q)(x) = B(x, q). \qquad (16.33)$$

From (16.29), (16.30), (16.32), and (16.33),

$$u \in \mathcal{A}^{-1}(\nabla \cdot \lambda), \qquad (16.34)$$

$$p \in \mathcal{B}^{-1}(\lambda); \qquad (16.35)$$

Combining now (16.31), (16.34) and (16.35), λ satisfies the following equation

$$\mathbf{0} \in -\nabla(\mathcal{A}^{-1}(\nabla \cdot \lambda)) + \mathcal{B}^{-1}(\lambda). \tag{16.36}$$

Problem (16.36) is a dual problem to problem (16.1). Concernng now algorithm (16.25) – (16.28), it can be shown (cf. e.g. [9]) that it is equivalent to a Douglas-Rachford operator splitting method applied to the time integration of the initial value problem: For $\lambda(0)$ given,

$$\mathbf{0} \in \frac{\partial \lambda}{\partial t} - \nabla(\mathcal{A}^{-1}(\nabla \cdot \lambda)) + \mathcal{B}^{-1}(\lambda), \tag{16.37}$$

where r becomes the integration time step δt.

16.7 Application to Example 1

In the particular case of Example 1 of Section 16.2, algorithm (16.25)–(16.28) becomes:

$$\lambda^0 \in (L^2(\Omega))^2, \quad u^{-1} \in H_0^1(\Omega) \text{ are given}; \tag{16.38}$$

for $n \geq 0$, assuming that λ^n and u^{n-1} are known, compute p^n, u^n as the solution of

$$rp^n + g\frac{p^n}{|p^n|} = r\nabla u^{n-1} + \lambda^n, \tag{16.39}$$

$$-(\mu + r)\nabla^2 u^n = \nabla \cdot \lambda^n - r\nabla \cdot p^n \text{ in } \Omega, \tag{16.40}$$

$$u^n = 0 \text{ on } \Gamma,$$

and update λ^n by

$$\lambda^{n+1} = \lambda^n + r(\nabla u^n - p^n). \tag{16.41}$$

Problem (16.40) is a classical Dirichlet problem. On the other hand, problem (16.39) requires some further discussion. It follows from (16.39) and from the positivity of r and g that:

$$p^n \text{ and } r\nabla u^{n-1} + \lambda^n \text{ are positively proportional.} \tag{16.42}$$

Taking, therefore, the scalar product of p^n with both sides of (16.39), we obtain from (16.42)

$$r|p^n|^2 + g|p^n| = \left|r\nabla u^{n-1} + \lambda^n\right| |p^n| \tag{16.43}$$

It follows then from (16.43) that

$$|p^n| = \begin{cases} \mathbf{0} \text{ if } g \geq |r\nabla u^{n-1} + \lambda^n|, \\ \frac{1}{r}(|r\nabla u^{n-1} + \lambda^n| - g) \text{ if } g < |r\nabla u^{n-1} + \lambda^n| \end{cases} . \tag{16.44}$$

Once $|p^n|$ is known from (16.44), we obtain p^n from (16.39).

16.8   Finite Element Implementation

To minimize technicalities we consider the case where Ω is a bounded polygonal subdomain of $\mathbb{R}^2$. Also, we shall consider in this section piecewise linear approximations only (piecewise quadratic approximations will be considered in Section 16.9, for the particular case associated with problem (16.3) from Section 16.2).

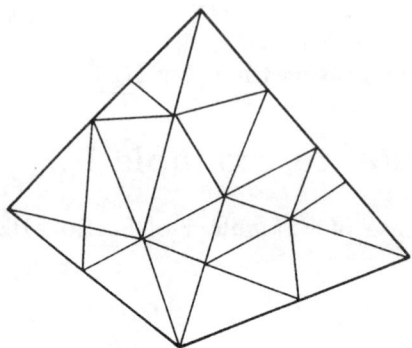

Figure 16.1: Triangulation of polygon.

Thus, let us consider a standard triangulation τ_h of Ω (such as that depicted in Figure 16.1) and denote by h the length of the largest edge(s) of τ_h. To approximate the boundary value problem (16.1) we use its variational formulation given in Section 16.3 (cf. (16.10)) and approximate $H^1(\Omega)$, $H_0^1(\Omega)$, $H_g^1(\Omega)$ by

$$V_h = \left\{ v_h | v_h \in C^0(\bar{\Omega}), \ v_h|_T \in P_1, \ \forall \, T \in \tau_h \right\}, \tag{16.45}$$

$$V_{oh} = \{ v_h | v_h \in V_h, \ v_h = 0 \text{ on } \Gamma \} = V_h \cap H_0^1(\Omega), \tag{16.46}$$

$$V_{gh} = \{ v_h | v_h \in V_h, \ v_h = g_h \text{ on } \Gamma \} ; \tag{16.47}$$

In (16.45), P_1 is the space of polynomials in x_1, x_2 of degree ≤ 1; (g_h is an approximation of g, the trace over Γ of a function $\tilde{g}_h$ belonging to V_h). We use the above spaces to approximate (16.1) by the following finite dimensional analog of (16.10): Find $u_h \in V_{gh}$ such that

$$\int_\Omega A(x)\nabla u_h \cdot \nabla v_h dx + \int_\Omega f(x, u_h)v_h dx + \int_\Omega B(x, \nabla u_h) \cdot \nabla v_h dx = 0,$$

$$\forall \, v_h \in V_{oh} ; \tag{16.48}$$

In practice, we should approximate the first and third integrals in (16.48) by

$$\sum_{T \in \tau_h} \text{meas.}(T) A(G_T) \nabla u_h \cdot \nabla u_h \text{ and}$$

$$\sum_{T \in \tau_h} \text{meas.}(T) B(G_T, \nabla u_h), \qquad (16.49)$$

respectively. In (16.49), meas.(T) is the area of triangle T and G_T the centroid of this triangle. (Here, we take advantage of the fact that ∇v_h is a piecewise constant vector in $\mathbb{R}^2$ over each triangle of τ_h. Similarly we should use the trapezoidal rule to approximate the second integral in (16.48). To solve problem (16.48) (or its variants obtained by numerical integration) we use the following discrete form of the multiplier algorithm (16.25)–(16.28) with

$$Q_h = \left\{ q_h | q_h \in L^2(\Omega) \times L^2(\Omega), q_h |_T = \text{const.}, \forall T \in \tau_h \right\}:$$

$$\lambda^0 \in Q_h, \ u_h^{-1} \in V_{gh} \text{ are given}; \qquad (16.50)$$

for $n \geq 0$, λ_h^n, u_h^{n-1} being given, compute p_h^n, u_h^n, λ_h^{n+1} as follows (we use (16.49) to simplify the presentation)

$$B(G_T, p_h^n |_T) + r p_h^n |_T = \left(r \nabla u_h^{n-1} + \lambda_h^n \right) |_T, \ \forall T \in \tau_h, \qquad (16.51)$$

$$\sum_{T \in \tau_h} \text{meas.}(T) A(G_T) \nabla u_h^n \cdot \nabla v_h + r \int_\Omega \nabla u_h^n \cdot \nabla v_h dx$$

$$+ \int_\Omega f(x, u_h^n) v_h dx = \int_\Omega (r p_h^n - \lambda_h^n) \cdot \nabla v_h dx, \qquad (16.52)$$

$$\forall v_h \in V_{oh}; \ u_h^n \in V_{gh}$$

$$\lambda_n^{n+1} = \lambda_h^n + r(\nabla u_h^n - p_h^n). \qquad (16.53)$$

The above algorithm deserves further comment:

1. Step (16.51) leads to a collection of 2-dimensional nonlinear problems (one for each triangle); since these problems are totally decoupled one from another they can be solved simultaneously, which in principle is appealing for a massively parallel approach. The above observation is the main justification for this approach in a parallel processing context.

2. Problem (16.52) is well suited to relaxation methods; this is particularly true if one uses large values of r (which is sometimes required by some applications).

Indeed, if f is smooth enough, conjugate gradient methods preconditioned by the coefficient matrix associated with the quadratic form $\int_\Omega |\nabla v_h|^2 dx$, or by an element-by-element approximation of it, can be interesting alternatives to relaxation methods. In the following Section 16.9, we discuss the solution of problem (16.3) in Section 16.2 when piecewise quadratic finite element spaces are used for approximation. Finally, in Section 16.10 we shall discuss the solution of a problem related to the Yang-Mills equations in high energy physics.

16.9 Application to Visco-Plastic Flow

Piecewise linear approximations, such as those discussed in Section 16.8, are ideally suited to the practical implementation of the multiplier algorithm (16.25)–(16.28). The main reason for this is clearly that ∇u and λ are approximated by piecewise constant functions. Indeed, if implemented properly, a piecewise quadratic approximation will provide the same advantages, in addition to a higher accuracy (for the same number of unknowns). We illustrate this point by considering the solution of the visco-plastic flow problem (16.3) from Section 16.2. We suppose again that Ω is a polygonal domain of $I\!R^2$ and we consider triangulations τ_h of Ω similar to those introduced in Section 16.8. Next, we approximate $H^1(\Omega)$, $H_0^1(\Omega)$ and $L^2(\Omega) \times L^2(\Omega)$ by

$$V_h = \left\{ v_h | v_h \in C^0(\bar\Omega),\ v_h|_T \in P_2,\ \forall\, T \in \tau_h \right\}, \tag{16.54}$$

$$V_{oh} = \{ v_h | v_h \in V_h,\ v_h = 0 \text{ on } \Gamma \} = V_h \cap H_0^1(\Omega), \tag{16.55}$$

$$Q_h = \left\{ q_h | q_h \in L^2(\Omega) \times L^2(\Omega),\ q_h|_T \in P_1 \times P_1,\ \forall\, T \in \tau_h \right\}; \tag{16.56}$$

In (16.54), P_2 is the (six-dimensional) space of polynomials in x_1, x_2 of degree ≤ 2. An element v_h of V_h is uniquely defined by the values it takes at the vertices and the mid-sides of the triangles of τ_h (see Figure 16.2). If $v_h \in V_h$ then $\nabla v_h \in Q_h$ and therefore, $\forall\, T \in \tau_h$, $\nabla v_h|_T$ is uniquely defined by the values it takes at the mid-side points of T (it is clear that ∇v_h is in general discontinuous at the mid-points of τ_h); the same property holds for the elements q_h of Q_h. To approximate (16.3) we shall use the following finite dimensional variational formulation. (The discretization of the nonlinear term

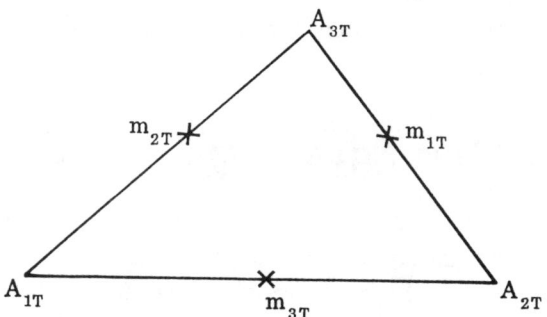

Figure 16.2: Vertex and mid-point values for quadratic triangle.

is based on the two-dimensional Simpson rule):

$$\mu \int_\Omega \nabla u_h \cdot \nabla v_h dx + g \sum_{T \in \tau_h} \frac{\text{meas.}(T)}{3} \sum_{i=1}^{3} \frac{\nabla u_h(m_{iT}) \cdot \nabla v_h(m_{iT})}{|\nabla u_h(m_{iT})|}$$

$$= \int_\Omega f v_h dx, \quad \forall\, v_h \in V_{oh}; \quad u_h \in V_{oh}. \tag{16.57}$$

The discrete analog of the multiplier algorithm (16.38) – (16.41), takes here the following form:

$$\lambda_h^0 \in Q_h, \; u_h^{-1} \in V_{oh} \text{ are given;} \tag{16.58}$$

then for $n \geq 0$, assuming that λ_h^n and u_h^{n-1} are known, we compute p_h^n, u_h^n and λ_h^{n+1} as follows:

$$r p_T^n \left(m_{iT} + g \frac{p_T^n(m_{iT})}{|p_T^n(m_{iT})|} \right) = r \nabla u_h^{n-1} + \lambda_h^n|_T(m_{iT}),$$

$$\forall\, i = 1, 2, 3, \forall\, T \in \tau_h \tag{16.59}$$

(we use the notation $p_T^n = p_h^n|_T$),

$$(\mu + r) \int_\Omega \nabla u_h^n \cdot \nabla v_h dx = \int_\Omega (r p_h^n - \lambda_h^n) \cdot \nabla v_h dx,$$

$$\forall\, v_h \in V_{oh}, \; u_h^n \in V_{oh}, \tag{16.60}$$

$$\lambda_h^{n+1} = \lambda_h^n + r(\nabla u_h^n - p_h^n). \tag{16.61}$$

Problem (16.60) is a classical approximate linear Dirichlet problem which can be solved by various iterative or direct methods. On the other hand, the discussion of Section 16.7 concerning the solution of problem (16.39) still applies for problem (16.59).

16.10   Yang-Mills Equations

An important application of the general methodology discussed in this paper is concerned with the solution of nonquadratic, nonconvex minimization problems, associated to the Skyrme model of the nucleon (see, e.g., [1] for the physical background, and [6] for existence results). From a mathematical point of view these problems are closely related to variational problems in finite elasticity and liquid crystal theory for which multiplier methods have been quite successful, as shown in e.g. [9]. A typical problem related to the Skyrme model is defined as follows: Find $u \in E$ such that

$$J(u) \le J(v), \quad \forall\, v \in E, \tag{16.62}$$

where, in (16.62), $u = \{u_k\}_{k=1}^4$, $v = \{v_k\}_{k=1}^4$ and $J(\cdot)$ and E are defined by

$$J(v) = \frac{\nu}{2} \int_{R^3} |\nabla v|^2\, dx + \frac{1}{2} \int_{R^3} |A(v)|^2\, dx, \tag{16.63}$$

with $\nu > 0$ and

$$|\nabla v|^2 = \sum_{k=1}^4 \sum_{i=1}^3 \left| \frac{\partial v_k}{\partial x_i} \right|^2, \tag{16.64}$$

$$|A(v)|^2 = \sum_{\substack{1 \le k,l \le 4 \\ k \ne l}} \sum_{\substack{1 \le i,j \le 3 \\ i \ne j}} \left| \frac{\partial v_k}{\partial x_i} \frac{\partial v_l}{\partial x_j} - \frac{\partial v_k}{\partial x_j} \frac{\partial v_l}{\partial x_i} \right|^2, \tag{16.65}$$

and by

$$E = \left\{ v | J(v) < +\infty\,, |v(x)| = 1 \text{ a.e.}, \int_{R^3} \det(v, \nabla v) dx = 2\pi^2 N \right\} \tag{16.66}$$

with $N \in \mathbb{Z}$. Actually, the integral relation in (16.66) implies that the extremity of vector $v(x)$ describes the unit sphere S^3 of $\mathbb{R}^4$ $|N|$ times as x describes $\mathbb{R}^3$. From various invariance properties of J and E, and from the geometrical interpretation of the integral relation in (16.66), we can assume (following [1]) that the solution of problem (16.62) satisfies

$$\lim_{|x| \to +\infty} u(x) = \{0,0,0,1\}. \tag{16.67}$$

In order to compute u we approximate problem (16.62) by: Find $u \in E_L$ such that

$$J_L(u) \leq J_L(v), \quad \forall \, v \in E_L, \tag{16.68}$$

where, in (16.68), $J_L(\cdot)$ and E_L are defined by

$$J_L(v) = \frac{\nu}{2} \int_{\Omega_L} |\nabla v|^2 \, dx + \frac{1}{2} \int_{\Omega_L} |A(v)|^2 \, dx, \tag{16.69}$$

with $\Omega_L = (-L, L)^3$, $(L > 0)$, and by

$$E_L = \{v \, | J_L(v) < +\infty \, , \, |v(x)| = 1 \quad \text{a.e.}, \tag{16.70}$$

$$\int_{\Omega_L} \det(v, \nabla v) dx = 2\pi^2 N, v = \{0, 0, 0, 1\} \text{ on } \partial \Omega_L \} \, ;$$

In practice L will have to be "sufficiently large". From a computational point of view it is convenient to map Ω_L onto $\Omega = (-1, +1)^3$ via the transformation

$$\xi = x/L. \tag{16.71}$$

Problem (16.68) then becomes (retaining some of the previous notations): Find $u \in E$ such that

$$J(u) \leq J(v), \forall \, v \in E, \tag{16.72}$$

where

$$J(v) = \frac{\nu}{2} \int_{\Omega} |\nabla v|^2 \, d\xi + \frac{1}{2L^2} \int_{\Omega} |A(v)|^2 \, d\xi, \tag{16.73}$$

$$E = \{v | J(v) < +\infty, \, |v(\xi)| = 1 \text{ a.e.},$$

$$\int_{\Omega} \det(v, \nabla v) d\xi = 2\pi^2 N, \, v = \{0, 0, 0, 1\} \text{ on } \partial \Omega \} . \tag{16.74}$$

Following the approach of [7] and [9], we associate with (16.72)–(16.74) a Lagrangian functional $\mathcal{L}_r$ defined by

$$\mathcal{L}_r(v, q; \mu) = \frac{\nu}{2} \int_{\Omega} |\nabla v|^2 \, d\xi + \frac{1}{2L^2} \int_{\Omega} B(q) d\xi$$

$$+ \frac{r}{2} \int_{\Omega} |\nabla v - q|^2 \, d\xi + \int_{\Omega} \mu \cdot (\nabla v - q) d\xi, \tag{16.75}$$

and the set

$$E_2 = \{v | v \in (H^1(\Omega))^4, \, |v(\xi)| = 1 \text{ a.e.},$$

$$v = \{0, 0, 0, 1\} \text{ on } (\Omega) \}, \tag{16.76}$$

where in (16.75) the functional $B = I\!\!R^2 \to I\!\!R$ is defined by

$$B(q) = \sum_{\substack{1 \le k,l \le r \\ k \ne l}} \sum_{\substack{1 \le i,j \le 3 \\ i \ne j}} |q_{ki}q_{lj} - q_{kj}q_{li}|^2 . \qquad (16.77)$$

To solve problem (16.72), we use the following variant of algorithm (16.25)–(16.28):

$$\lambda^0 \in (L^2(\Omega))^{12} , \quad u^{-1} \in (H^1(\Omega))^4 \text{ are given;} \qquad (16.78)$$

then for $n \ge 0$, assuming that λ^n, u^{n-1} are known compute p^n, u^n and λ^{n+1} as follows: Find $p^n \in E_1^n$ such that

$$\mathcal{L}_r(u^{n-1}, p^n, \lambda^n) \le \mathcal{L}_r(u^{n-1}, q; \lambda^n), \quad \forall\, q \in E_1^n , \qquad (16.79)$$

where

$$E_1^n = \left\{ q | q \in (L^2(\Omega))^2 , \int_\Omega \det(u^{n-1}, q) d\xi = 2\pi^2 N \right\} , \qquad (16.80)$$

and: Find $u \in E_2$ such that

$$\mathcal{L}_r(u^n, p^n, \lambda^n) \le \mathcal{L}_r(v, p^n; \lambda^n), \quad \forall\, v \in E_2 , \qquad (16.81)$$

$$\lambda^{n+1} = \lambda^n + r(\nabla u^n - p^n) . \qquad (16.82)$$

Problem (16.80) is very similar to the liquid crystal problem whose numerical solution is discussed in [2,3,4,9]; indeed the operator splitting methods and the relaxation methods discussed in the four above references can be applied to the solution of problem (16.81).

Solving problem (16.81) is apparently more complicated. In fact the solution p^n of (16.79) has to satisfy the following system of equations:

$$\frac{1}{2L^2} B'(p^n) + rp^n + z^n \frac{\partial}{\partial q} \det(u^{n-1}, p^n) = \lambda^n + r\nabla u^n , \qquad (16.83)$$

$$\int_\Omega \det(u^{n-1}, p^n) d\xi = 2\pi^2 N . \qquad (16.84)$$

In (16.83), $z^n (\in I\!\!R)$ is a Lagrange multiplier associated with relation (16.84). Solving the above system (16.84) is quite easy by the Newton's/bordering algorithm described in [13]. Actually, if the problem is discretized by piecewise linear finite element approximations (like those in Section 8), then p^n, λ^n and ∇u^n will be piecewise constant, implying that the solution can be achieved simultaneously, element-by-element, making it massively parallelizable. The numerical solution on parallel computers of the Skyrme problem, by the methods described above is presently being investigated by E. Dean, H. B. Keller, E. Van de Velde and the author.

16.11 Conclusion

We have presented in this chapter a multiplier/element-by-element method and discussed its application to problems from mechanics and physics. Among the significant features of this approach we mention in particular:

1. It decouples the nonlinearity in the differential operators.

2. Hard nonlinearities are "localized" and reduced to small dimensional decoupled nonlinear problems, which can be solved in parallel.

3. This methodology is very robust, mainly because it is very implicit with respect to the nonlinearity; also the additional term associated with r plays, at least locally, the role of a smoother and of a convexifier, implying good convergence properties of the algorithms discussed in this paper.

4. As shown in [7,9,10,11] the methodology described in this paper applies also to 4th order nonlinear problems.

5. It is well suited to finite element implementations.

The main disadvantage of the multiplier/element by element method discussed here is the extra unknowns, p and λ, that it introduces. However, there are many applications for which this is offset by the advantages noted above (as shown for example in [7,9]).

Acknowledgment

We thank M. J. Esteban, M. Fortin, H. B. Keller, P. Le Tallec, and E. Van de Velde for helpful comments and suggestions. We also thank J. Cassidy for her assistance.

References

[1] Aitchison, I. J. R., "The Skyrme Model of the Nucleon," Report 2/88, University of Oxford, Department of Theoretical Physics, 1988.

[2] Cohen, R., S. Y. Lin and M. Luskin, "Relaxation and Gradient Methods for Molecular Orientation in Liquid Crystals," Report UMSI 88/97, University of Minnesota, Supercomputer Institute, October 1988.

[3] Dean, E., R. Glowinski and C. H. Li, "Applications of Operator Splitting Methods to the Numerical Solution of Nonlinear Problems in Continuum Mechanics and Physics," in *Mathematics Applied to Science,* J. Goldstein, S. Rosencrans, G. Sod (eds.), Academic Press, Boston, 13–64, 1988.

[4] Dean, E., R. Glowinski and C. H. Li, "Numerical Solution of Parabolic Problems in High Dimensions," in *Transactions of the Fifth Army Conference on Applied Mathematics and Computing,* ARO Report 88–1, 207–285, 1988.

[5] Duvaut, G. and J. L. Lions, *Les Inequations en Mecanique et en Physique,* Dunod, Paris, 1972.

[6] Esteban, M. J., "A Direct Variational Approach to Skyrme's Model for Meson Fields," *Commun. Math. Phys.,* 105, 571–591, 1986.

[7] Fortin, M. and R. Glowinski, *Augmented Lagrangians,* North-Holland, Amsterdam, 1983.

[8] Glowinski, R., *Numerical Methods for Nonlinear Variational Problems,* Springer, New York, 1984.

[9] Glowinski, R. and P. Le Tallec, *Applications of Augmented Lagrangians and Operator Splittings in Nonlinear Mechanics,* SIAM, Phildelphia, 1989.

[10] Glowinski, R., J. L. Lions and R. Tremolieres, *Numerical Analysis of Variational Inequalities,* North-Holland, Amsterdam, 1981.

[11] Glowinski, R., D. Marini and M. Vidrascu, "Finite Element Approximations and Iterative Solutions of a Fourth-Order Elliptic Variational Inequality," *IMA Journal of Numerical Analysis,* 4, 127–167, 1984.

[12] Lions, P. L. and B. Mercier, "Splitting Algorithms for the Sum of Two Operators," *SIAM J. Num. Anal.,* 16, 964–979, 1979.

[13] Keller, H. B., *Numerical Methods in Bifurcation Problems,* Springer-Verlag, Berlin, 1987.

Chapter 17

Exploring Parallel Algorithms Having No Serial Analogues

*Robert Hiromoto**

17.1   Introduction

The ordering of data acquisitions in many computational problems is an artifact of algorithms developed for serial computers. Often these serial algorithms are highly parallel and thus can be mapped directly onto a parallel processing system. This technique, however, does not fully exploit the additional opportunities provided by the system's parallelism. The design of optimal parallel algorithms requires new and different techniques and insights. Unfortunately, parallel performance can be degraded even with optimal parallel algorithms that preserve the ordering of data access (that is, the data dependences between different computational code blocks) by synchronization primitives such as locks, events, and barriers. Furthermore, additional decreases in performance are introduced by the various hardware/software components integrated to coordinate the particular parallel processing system.

With the introduction of parallel processing systems, a variety of programming options have become available in the modeling of physical simulation problems. Likewise, older more computationally intensive methods, such as

*Computing And Communications Division, Los Alamos National Laboratory.

the Monte Carlo method, may become more attractive within the framework of a parallel processing environment. Exploring and evaluating old and new parallel techniques not only enhances their further development, but also may provide insight that leads to more efficient algorithms. A chaotic iterative technique is one interesting alternative to the standard parallel algorithms that preserve the ordering of data access. In a chaotic iterative algorithm, the synchronous access of data between parallel execution streams is ignored. The values of new iterates are computed by using data available at the instant data requests are made by each competing process. A disadvantage is that the data that are accessed may lag behind the current iteration by an iteration step α. The use of "stale" data values may have serious consequences on the rate of convergence guaranteed by standard iterative methods.

Chaotic schemes are useful, therefore, if the rate of convergence is reasonably insensitive to the delay α. The delay α depends on how the iterative scheme interacts with itself and with the parallel processing system. The asynchronous approach still is interesting and attractive for several reasons. First, the technique minimizes (or eliminates) the overhead incurred by synchronizing processes. Second, it increases the parallelism by reducing the amount of time that processes wait on synchronizing events. Third, the chaotic techniques provide a simple means of simulating nonl-local interacting fields. These interactions are characterized by events separated by time delays that are difficult to model in sequential algorithms. Although numerous stability issues can arise, the chaotic approach still may be useful in modeling such computational problems. Fourth, it provides the user with an easier programming style for writing parallel codes. Fifth, as chaotic schemes are system dependent, they may provide a useful means to evaluate the characteristics of a parallel processing system. Finally, the insights gained from experimenting with these methods to determine what serial analogues exist, might possibly lead to improved serial and parallel algorithms.

Over the past few years, B. R. Wienke, R. G. Brickner, and R. Hiromoto have reported on chaotic techniques for the deterministic transport algorithm S_n [2,12,13]. This work is focused on the traditional, double iteration scheme commonly referred to as the inner-outer scheme. It was motivated by the structure of the S_n algorithm and the availability of experimental parallel processing systems. The computational nature of the S_n algorithm uses iterative techniques [4] in solving the linear Boltzmann equation. Various schemes have been implemented on the Denelcor HEP [9], the Intel hypercube [6], and more recently on the Encore Multimax [5]. The different algorithms were also implemented in a production transport code called ESN [10,11], which is a one-dimensional, time-dependent, Lagrangian S_n module. The parallelizing techniques employ both ordered and chaotic inner energy-group sweeps as described below.

This chapter compares the results of these chaotic schemes with a corresponding deterministic parallel scheme. Using this comparison, a possible serial analogue to the chaotic schemes is offered that leads to an improved rate of convergence.

17.2 Description

The S_n algorithm is a deterministic particle transport algorithm that solves the linear Boltzmann transport equation. Solutions to this equation are important in the study of dynamic particle systems in which the aggregate behavior may be affected by collisions, absorptions, or the infusion of external particles. The computational form of the equation (referred to as the multigroup, discrete ordinates equation) is derived by the discretization of space and the partitioning of the particle flux into discrete energy groups with discrete angular directions. This iterative algorithm is divided into an inner-outer loop, where each loop might be repeated many times. The iteration ranges over the energy and angular domains that are designated as G and M, respectively.

In the inner loop, each of the energy groups, g, iterates over the corresponding angular flux sources in each of m directions. The iterative scheme uses either old iterates or a mixture of old and new iterates to compute the new iterates. This choice depends on the iterative scheme selected. In either case, the angular flux of each energy group is checked, and when all the group fluxes have converged, the program proceeds to the outer loop. However, if an energy group does not converge, that particular group is reiterated in the inner loop.

At the outer loop, the flux is integrated over all groups and then compared with the flux from the previous outer-loop cycle. The program halts if this scalar flux is converged; otherwise, the program returns to the inner loop and cycles between the inner and outer loops. Figure 17.1 illustrates a flow diagram of the inner-outer schemes. For future reference, the left side of Figure 17.1 represents the flow of the parallel execution, whereas the right side indicates the sequential flow and defines the various numbered execution threads. The results of the iterative schemes described below will be used to compare the effects of synchronous and asynchronous (chaotic) parallel multigroup techniques.

A compact form of the multigroup, discrete ordinates equation can be

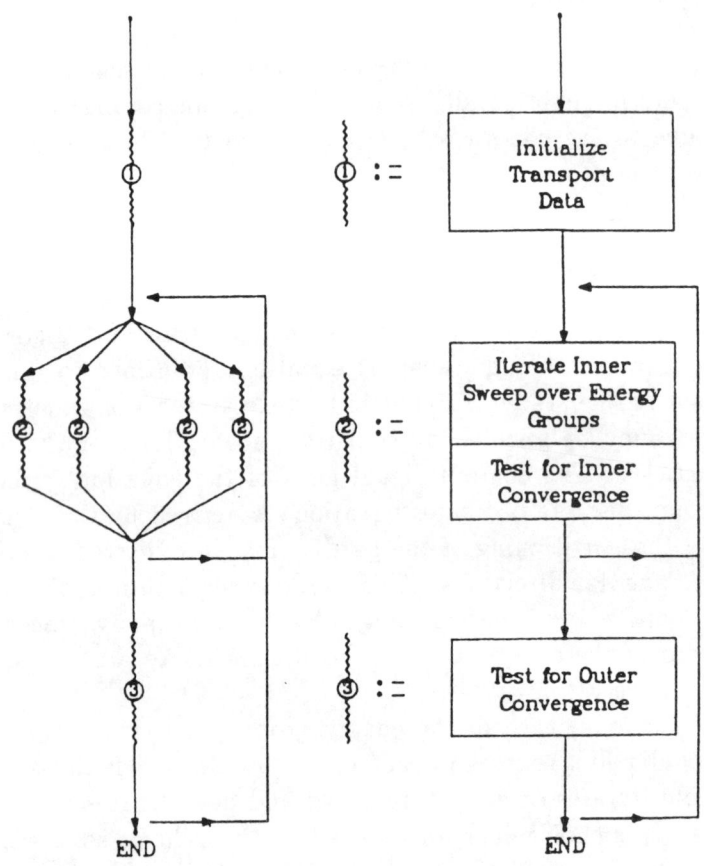

Figure 17.1: Flow diagram for the deterministic transport algorithm S_n.

written as (Carlson [3])

$$v_g^{-1}\frac{\partial \phi_{gm}}{\partial t} + \boldsymbol{\Omega} \cdot \boldsymbol{\nabla} \phi_{gm} + \sigma_g \, \phi_{gm} = \sum_{h=1}^{G} \sum_{l=1}^{M-1} \sigma_{hg}^l \phi_h^l P_l(\Omega_m) + Q_{gm}, \quad (17.1)$$

with ϕ the particle flux, t the time, Q the external source, v the velocity, σ the transport cross section, and $\boldsymbol{\Omega}$ a unit vector in the direction of particle travel. The indices g and m are the indices on the discrete energy and angular domains and are given as follows:

$$1 \ \leq \ g \ \leq \ G \qquad \text{(energy)}$$

$$1 \ \leq \ m \ \leq \ M \qquad \text{(angle)} \ .$$

The subscript g in Eq. (17.1) denotes appropriate group averaged quantities. The multigroup angular fluxes are functions of particle position, direction, energy, and time. The two quantities ϕ_h^l and σ_{hg}^l are defined by a Legendre expansion. The differential scattering-group cross sections σ_{hg}^l satisfy the normalization

$$\sigma_g^l = \sum_{h=1}^{G} \sigma_{gh}^l \,. \tag{17.2}$$

Equation (17.1) is an integro-differential equation that is solved by successively iterating the angular flux occurring in the first term of the right-hand side [1,7,8]. Iterations on Eq. (17.1) with $h = g$ are the within-group or inner iterations, whereas iterations with $h \neq g$ represent outer iterations. It is convenient to rewrite Eq. (17.1) in the following operator form

$$(L + \Sigma)\Phi = (S + U + D)\Phi + Q \,, \tag{17.3}$$

where Φ is the flux and L, Σ, S, U, D and Q are the streaming, collision, self-scatter, upscatter, downscatter, and external source operators, respectively. The upscatter and downscatter operators refer to energy transfers between higher or lower energy groups. The self-scatter operator implies no energy transfer.

A typical inner-outer iterative scheme is implemented as follows. Upscatter and downscatter operators couple different energy groups (outer), whereas self-scatter only couples within-group fluxes (inner). Using an iteration index i, the outer scheme takes the symbolic form

$$(L + \Sigma)\Phi^{i+1} = (S + D)\Phi^{i+1} + U\Phi^i + Q \,. \tag{17.4}$$

The upscatter source is computed from the previous iterate $U\Phi^i$, and the external source Q is constant. For inner iterations in the source sweeps from higher to lower energy groups, it is convenient to define a fixed effective source QQ^i as

$$QQ^i = D\Phi^{i+1} + U\Phi^i + Q \,. \tag{17.5}$$

Here the downscatter source $D\Phi^{i+1}$ involves only contributions from higher groups. Having obtained QQ^i, one then solves the within-group equation in the inner strategy

$$L\Phi^{i+1,j+1} + \Sigma \Phi^{i+1,j+1} = S\Phi^{i+1,j} + QQ^i \,, \tag{17.6}$$

with j denoting the inner-iteration cycle, in analogy to the outer index i. During this computational cycle, Eq. (17.6) is iterated until the fluxes converge. Each outer iteration thus involves one pass through all energy groups in solving Eq. (17.6). Convergence is again tested at the end of the outer cycle, and

QQ^i is updated for the next iteration if necessary. This dual iterative strategy is called the inner-outer sweep, with spectral radii and convergence properties determined by the norms of the appropriate iteration matrices, $(L + \Sigma)^{-1} S$ in the inner case and $(L + \Sigma - S - D)^{-1} U$ in the outer case.

Using the standard operator form, the two sequential iterative schemes, termed TPMG (multigroup) and TPS (standard), respectively, take on the operational forms

$$(L + \Sigma)\Phi^{i+1,j+1} = S\Phi^{i+1,j} + (D + U)\Phi^i + Q \qquad (17.7)$$

and

$$(L + \Sigma)\Phi^{i+1,j+1} = S\Phi^{i+1,j} + D\Phi^{i+1} + U\Phi^i + Q . \qquad (17.8)$$

TPMG is a block Jacobi-like scheme. The outer-flux source terms, $D\Phi^i$ and $U\Phi^i$, are updated after *all* within-group fluxes have converged. Hence, a direct mapping from the sequential scheme to a parallel TPMG scheme is easily obtained by placing a barrier synchronization point before the outer-loop convergence test. The parallel version of TPMG also preserves the sequential access pattern of converged-group flux iterates; therefore, it is a deterministic parallel scheme. TPS is a block Gauss-Seidel-like scheme in the downscatter source $D\Phi^{i+1,j}$. When a within-group flux converges, its contributions to $D\Phi^{i+1,j}$ are updated. As the inner-loop scheduling suggests, the sequential TPS scheme imposes an artificial constraint that allows outer-flux iterates to be passed in only one direction (downscatter). TPMG, of course, is restricted even more because the use of new outer-source iterates are only allowed from one outer-loop iteration to the next.

Because of the inner-loop scheduling of energy groups and the propagation of downscatter sources, the mapping of the sequential TPS scheme directly to a parallel TPS scheme is impractical. Rather than devise an entirely new algorithm for parallel execution, a simple alternative approach is taken, which ignores the sequentiality imposed by TPS. This parallel scheme allows the use of converged upscatter and downscatter sources by both higher and lower energy groups when executing the inner-iteration loop. That is, $U\Phi^i$ effectively becomes $U\Phi^{i+1,j}$ in Eq. (17.8). The coupling of different energy groups is done by not synchronizing parallel processes executing in the inner loop. However, a synchronization barrier is used to coordinate the convergence of all within-group fluxes, before the outer-loop (global) convergence check is performed.

In both the parallel TPMG and TPS schemes, it is important to note that the inner loops are synchronized so that all inner loops are finished and all flux iterates converged, before the next outer cycle begins. Beginning with these two sequential and parallel algorithms, other chaotic parallel schemes are explored in the following sections.

17.3   Results and Analysis

The test problem is a 16-group, isotropic boundary source of photons, which are Planckian distributed in energy at T = 20 keV, incident on a sphere of 6 to 10 mean free paths in thickness. The scattering ratio varies from 0.6 to 0.85, and all energy groups are Compton coupled. An S_4 quadrature is used with a convergence criterion of 10^{-3} for both inner and outer loops. No distributed source is present. The problem was run both in serial and in parallel modes. A dynamic scheduling technique was implemented, allowing group sweeps to be executed in parallel using varying numbers of processes (1 to 16). Specific statistics for timing and iteration counts will be presented graphically.

The parallel execution profiles of TPMG and TPS for the HEP and the Encore are summarized in Figure 17.2. The total inner-iteration count measured on the Encore is also given. Unfortunately, no data are available for the HEP. Note that the total inner-iteration count for TPMG is constant, independent of the number of processes executed in parallel. This should be of no surprise because the parallel algorithm is Jacobi-like and maps directly back to the serial algorithm. In TPS, on the other hand, the rate of convergence varies as a function of the number of processes. Several reasons may explain the degrading performance of the chaotic TPS scheme. As an example, the chaotic behavior of TPS may introduce large and abrupt changes in the terms of the outer source as seen by an iterative process. This may happen if the process updates its flux iterates using outer sources that are α steps behind those sources present at the end of the current iteration step j. If the difference between the old-source and new-source terms fluctuates enough, the test for convergence fails. For this case, the inner sweeps might be thought of as converging on "false" iterates and, therefore, require additional inner-outer iterative cycles. Assuming that the effects of large outer-source fluctuations may be responsible for the observed behavior of TPS, a change to TPS that addresses this apparent problem should improve the measured results.

To test this conjecture, a second chaotic scheme based on TPS was designed and implemented. This scheme, referred to as TPCC (chaotic convergence), updates a group's outer-flux contributions after each cycle of an inner loop. In contrast to TPS, a group's flux need not converge before it is updated. Because of this, TPCC is more chaotic and has the advantage of updating the outer source more frequently, with possibly smaller variations in the evolving source terms. Because TPCC is more chaotic, it does seem reasonable to ask if a serious bottleneck arises when the data memory access pattern is more asynchronous.

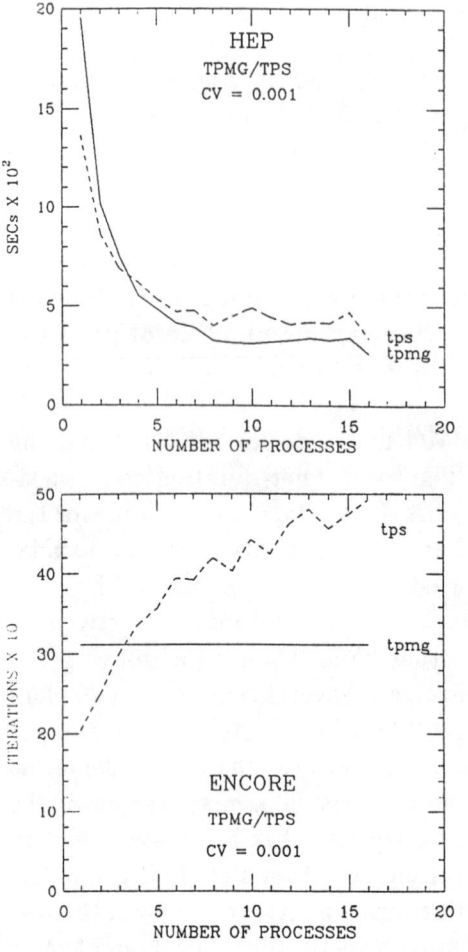

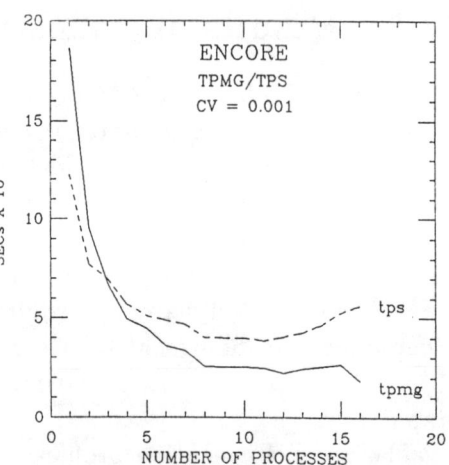

Figure 17.2: HEP and ENCORE parallel execution profiles for (a), (b) TPMG and TPS and ENCORE iteration count (c) with convergence criterion (CV) of 10^{-3}.

In operator form TPCC is given by

$$(L + \Sigma)\Phi^{i+1,j+1} = S\Phi^{i+1,j} + (\overline{D+U})\,\Phi^{i+1} + Q, \qquad (17.9)$$

where $(\overline{D+U})\Phi^{i+1}$ represents the outer source composed of source updates per iteration j per group.

The parallel execution times of the three schemes TPMG, TPS, and TPCC for the HEP and the Encore systems are shown in Figure 17.3. The total number of inner iterations required for outer convergence are given in Figure 17.4. The inner iteration counts for TPS on the HEP are not given because they are unavailable. As shown in these graphs, TPCC performs substantially better than TPS or TPMG.

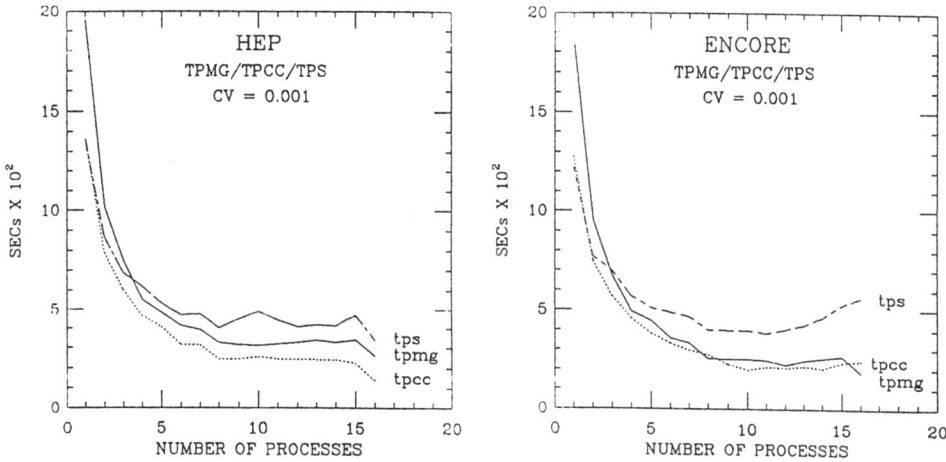

Figure 17.3: HEP and ENCORE parallel execution profiles for TPMG, TPCC, and TPS

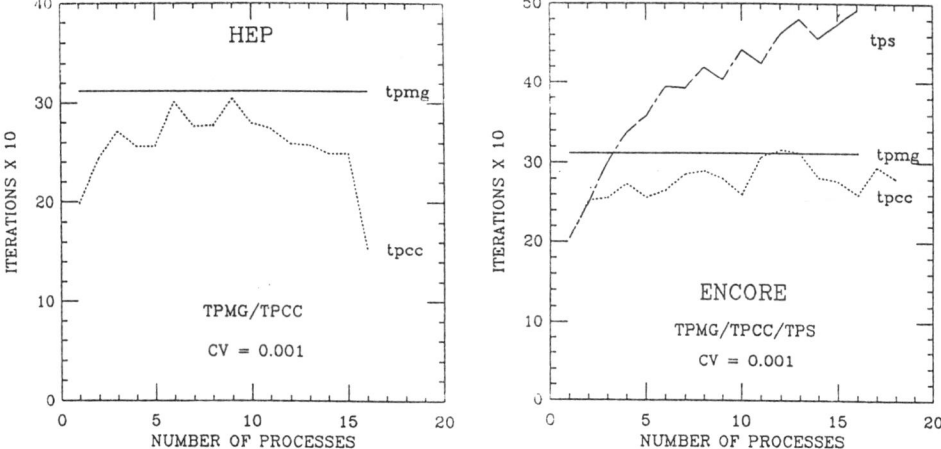

Figure 17.4: Total number of inner iterations for outer convergence using TPMG and TPCC on the HEP.

These results reflect not only the effects of the different chaotic algorithms but also the differences in the two parallel architectures. On the HEP, memory is accessed through a high-speed-memory switching network. A separate, pipelined data-memory access unit prevents processes, requesting operands from data memory, from blocking other processes ready to execute. The non-blocking of ready processes allows their execution to overlap (or partially hide) the memory access time delays. The outer-source values, shared by all parallel processes, are updated and available as soon as any process is ready to execute.

The Encore, on the other hand, uses local cache memory and a single high-speed bus for accessing shared memory. The Encore processes are blocked if a shared cache location is tagged "old" by the operating system. The Encore operating system suspends (blocks) the process requiring an operand from a tagged cache location, updates the tagged location in cache with the new value, and then restarts the suspended process.

Ignoring the differences in execution rates between the two parallel processing systems, the parallel profiles of the three schemes are surprisingly similar for the HEP and the Encore.

17.4   TPCS as a Serial Analogue

TPCC was an attempt to minimize the potentially large outer-source changes apparently induced by the chaotic TPS scheme. Based on the results given above, this problem appears to be solved, and the increase in the chaotic behavior of TPCC does not appear to degrade its performance. To further explore the effects of the outer-source on the rate of convergence, a slight change was made to TPCC. In this modified scheme, referred to as TPCS (chaotic single-pass), the inner loop is replaced by a single iteration sweep executed in parallel by all energy groups; in other words, for the inner-loop index $j = 1$. No inner convergence test is made on any of the group fluxes. For each group, a process iterates the flux for a single sweep and then updates the results. After completing a single sweep through the inner loop, processes synchronize before proceeding to the outer-iteration step and convergence check. If outer convergence is not satisfied, another single-pass inner sweep is executed for all groups. Otherwise, the execution is stopped. There are two advantages to the TPCS scheme. The first advantage is that the upscatter and downscatter fluxes cannot lag behind the current iteration by more than one step (that is, $\alpha = 1$). And the second advantage is that the fluctuations in the outer-source should be minimized further.

The results for both the Encore and the HEP are summarized in Figure 17.5. Again the data for the total number of inner iterations for the HEP are unavailable. TPCS performs surprisingly well. What is even more surprising is the sequential performance of TPCS. As a sequential scheme, the single-pass inner-iteration loop still passes only downscatter flux iterates for use by the remaining (lower) energy groups. However, regarding i (the outer-iteration index) as the current iteration through the single-pass inner loop, the ith single-pass iterations for all groups are based on the upscatter source iterates, updated from the previous $(i-1)$ iteration. In other words, the latest upscatter source iterates are available after each complete inner-

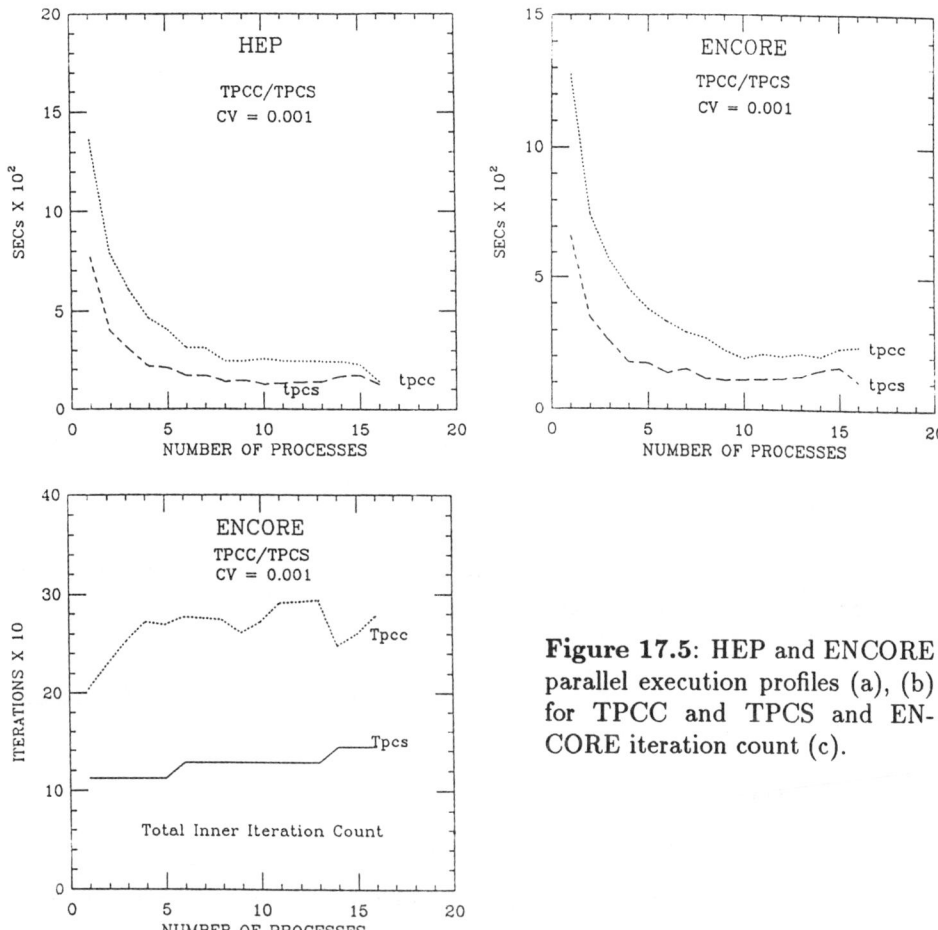

Figure 17.5: HEP and ENCORE parallel execution profiles (a), (b) for TPCC and TPCS and EN-CORE iteration count (c).

iteration loop. This sequential scheme more closely resembles the updating of outer-sources done chaotically.

17.5 Concluding Remarks

TPCS resulted from a series of numerical experiments that provided an indirect analysis of the behavior of one deterministic and two chaotic iterative schemes. The apparent iterative properties that characterize the different schemes are the frequency of outer-source updating, the fluctuations of outer-source terms during an inner iteration, and the parameter α that measures the delay of the chaotically evolving outer-source terms. The chaotic inner iterations are controlled and bounded by a barrier synchronization at the outer loop. The test problem does not include the case with distributed sources. For this reason, the generality of TPCS is not clear. However, TPCS does provide an example of a chaotic algorithm, developed through a series of numerical experiments, which has a serial scheme that appears to embody analogous, chaotic features.

References

[1] Alcouffe R. E., E. W. Larsen, W. F. Miller, and B. R. Wienke, "Computational Efficiency of Numerical Methods for the Multigroup Discrete Ordinate Neutron Transport Equations: The Slab Geometry Case," *Nuc. Sci. Eng.*, 71, 111-127, 1979.

[2] Brickner R. G., R. E. Hiromoto, and B. R. Wienke, "Parallel Iterative Transport Algorithms and Comparative Performance on Distributed and Common Memory Systems," Los Alamos National Laboratory Report: LA-UR-87-2163.

[3] Carlson, B. G., "A Method of Characteristics and Other Improvements in Solution Methods for the Transport Equation," *Nuc. Sci. Eng.*, 61, 408-425, 1976.

[4] Carlson, B. G., and K. D. Lathrop, *Computing Methods In Reactor Physics*, Gordon and Breach, New York, 1968.

[5] Encore Computer Corp., *Multimax Technical Manual*, 1986.

[6] Intel Scientific Computers, Inc., *iPSC System Overview*, 1986.

[7] Lathrop, K. D., "Spatial Differencing of the Transport Equation: Positivity vs. Accuracy," *J. Comp. Phys.*, 4, 475–298, 1969.

[8] Reed, W. H., "The Effectiveness of Acceleration Techniques for Iterative Methods in Transport Theory," *Nuc. Sci. Eng.*, 45, 245–254, 1971.

[9] Smith, B. J., "A Pipelined, Shared Resource MIMD Computer," Proceedings of the 1978 International Conference on Parallel Processing, Bellaire, MI, 6–8, August 1978.

[10] Wienke, B. R., "ESN," *Nuc. Sci. Eng.*, 81, 302, 1982.

[11] Wienke, B. R., "ESN: One Dimensional S_n Transport Module for Electrons," *J. Quant. spect. Rad. Trans.*, 28, 311-326, 1982.

[12] Wienke B. R., and R. E. Hiromoto, "Parallel S_n Iterative Schemes," *Nuc. Sci. Eng.*, 90, 116–128, 1985.

[13] Wienke, B. R., and R. E. Hiromoto, "Chaotic Iteration and Parallel Divergence," *Parallel Comp.*, 85, 205–210, 1985.

Chapter 18

Application of Vectorization and Microtasking for Reservoir Simulation

*Amin Pashapour,[*] Gary A. Pope,[†] Kamy Sepehrnoori,[†]
and Gene Shiles[‡]*

18.1 Introduction

In this chapter we give the results of our research on vectorization and Microtasking of a reservoir simulation code. By reservoir simulator we mean a numerical model of the permeable medium in which fluid displacement is occurring. In general, this requires numerical solution of a set of coupled, non-linear partial differential equations describing complex physical processes in a three-dimensional domain. The most complex simulators are those used for enhanced oil recovery. Enhanced oil recovery methods involve the injection of chemicals, solvents or heat into the underground rock strata containing the oil to supplement the natural energy of these oil reservoirs and to increase the displacement of the oil trapped in the pores of the reservoir rock. Those methods using chemicals such as detergents, polymers and caustic are called chemical flooding techniques. Chemical flooding processes involve multiphase flow of a large number of components that interact with the porous medium and that in other ways lead to extreme nonlinearities in the partial differential

[*]Digital Equipment Corporation, Marlboro, MA 01725-9103.
[†]Department of Petroleum Engineering, The University of Texas, Austin, TX.
[‡]Cray Research Inc., Mendota Heights, MN 55120.

equations. Thus, the solution of this type of problem is difficult even when compared to other reservoir simulators, such as those used to model water-flooding. The partial differential equations are discretized by using a finite difference or finite element method, which in either case yields a system of algebraic equations. This system of equations must be solved at each time level by means of a solution algorithm.

Because chemical flooding is a process whereby a small amount of an oil-mobilizing chemical is injected into the formation, mixing and crossflow phenomena are especially important to its performance and need to be accurately modelled. This mixing is characterized by dispersion in both the longitudinal and transverse directions. Interaction of dispersion and phase behavior may cause the chemical slug to lose its effectiveness through several mechanisms, e.g. phase trapping. In other cases dispersion may either enhance the oil recovery efficiency by increasing the sweep efficiency, or it may lead to early breakdown of the process. In numerical simulation of these and other processes, however, the artificial numerical dispersion can further smear concentration fronts by increasing the level of dispersion, resulting in inaccurate prediction of both oil recovery and chemical transport. For this and other reasons, a relatively fine mesh is required to simulate chemical flooding. Also, small time steps are required because of the high degree of nonlinearity of the equations. Thus, to simulate large volumes typical of oil reservoirs, a very large amount of computer time as well as a large amount of storage is required. Only supercomputers can handle these large problems, and even then a highly optimized code is essential for practical and affordable simulations.

18.2   Chemical Flooding Simulator

The program we use at The University of Texas to simulate chemical flooding is a three-dimensional finite-difference program called UTCHEM. In this simulator, the material balance equations are solved for up to nineteen components: water, oil, surfactant, polymer, anions, divalent cations, cosurfactant 1, cosurfactant 2, three tracers, sodium dichromate, thiourea, trivalent chromium, gel, hydrogen, carbon, and organic acid species. These components may form up to three phases, aqueous, oleic and microemulsion, depending on the relative amounts and effective salinity of the phase environment.

The governing component conservation and overall material balance (pressure) equations are solved by a finite-difference approximation of the spatial derivatives and a forward-difference approximation of the time derivatives. First, the pressure equation is solved implicitly using explicit dating of concentration-dependent terms. Then the total component concentrations are solved explicitly. The phase behavior, saturations, and all fluid properties are then computed using various physical property models as a function of the total concentrations. The major physical phenomena modelled in UTCHEM are density, viscosity, phase behavior, dispersion, adsorption, interfacial tension, relative permeability, capillary pressure, capillary trapping of phases, cation exchange, polymer properties, gel properties, and various chemical reactions with the crude oil and the minerals in the oil-bearing rock.

Material balance equations for the isothermal flow of chemical species in three-dimensional permeable media are given in Datta Gupta *et al.* [2], Saad *et al.* [6], and Pope *et al.* [5]:

$$\frac{\partial}{\partial t}\left[\phi\tilde{c}_i\left(1+c_i^0\Delta P\right)\right]$$

$$+\frac{\partial}{\partial x}\left\{\sum_{j=1}^{n_p}\left(1+c_i^0\Delta P\right)\right.$$

$$\times\left[c_{ij}u_{xj}-\phi S_j\left(K_{xxij}\frac{\partial c_{ij}}{\partial x}+K_{xyij}\frac{\partial c_{ij}}{\partial y}+K_{xzij}\frac{\partial c_{ij}}{\partial z}\right)\right]\right\}$$

$$+\frac{\partial}{\partial y}\left\{\sum_{j=1}^{n_p}\left(1+c_i^0\Delta P\right)\right.$$

$$\times\left[c_{ij}u_{yj}-\phi S_j\left(K_{yxij}\frac{\partial c_{ij}}{\partial x}+K_{yyij}\frac{\partial c_{ij}}{\partial y}+K_{yzij}\frac{\partial c_{ij}}{\partial z}\right)\right]$$

$$+\frac{\partial}{\partial z}\left\{\sum_{j=1}^{n_p}\left(1+c_i^0\Delta P\right)\right.$$

$$\times\left[c_{ij}u_{zj}-\phi S_j\left(K_{zxij}\frac{\partial c_{ij}}{\partial x}+K_{zyij}\frac{\partial c_{ij}}{\partial y}+K_{zzij}\frac{\partial c_{ij}}{\partial z}\right)\right]\right\}$$

$$= Q_i \qquad i = 1,\ldots,n_c \tag{18.1}$$

where $\Delta P = P_j - P_R$ and

$$\tilde{c}_i = \left(1-\sum_{i=1}^{n_c}\bar{c}_i\right)\sum_{j=1}^{n_p}\left(c_{ij}S_j\right)+\bar{c}_i \tag{18.2}$$

The overall material balance equation is obtained by summing the above equations over all components.

$$\phi c_t \frac{\partial P}{\partial t} + \frac{\partial}{\partial x} \sum_{j=1}^{n_p} \left[u_{xj} \sum_{i=1}^{n_c} \left(1 + c_i^0 \Delta P\right) c_{ij} \right]$$

$$+ \frac{\partial}{\partial y} \sum_{j=1}^{n_p} \left[u_{yj} \sum_{i=1}^{n_c} \left(1 + c_i^0 \Delta P\right) c_{ij} \right]$$

$$+ \frac{\partial}{\partial z} \sum_{j=1}^{n_p} \left[u_{zj} \sum_{i=1}^{n_c} \left(1 + c_i^0 \Delta P\right) c_{ij} \right] = \sum_{i=1}^{n_c} Q_i \qquad (18.3)$$

where

$$c_t = c_f + \sum_{i=1}^{n_c} c_i^0 \, \tilde{c}_i \qquad (18.4)$$

The pressure equation is obtained from Equation (18.3) by substituting Darcy's law for multiphase flow for the velocity terms and using the definition of capillary pressure to write the equation in terms of a single phase pressure. The pressure equation is differenced and then solved implicitly in UTCHEM. The above species conservation equations are solved by explicit finite difference with dispersion control as given in Datta Gupta et $al.$ [2] and Saad et $al.$ [6].

The discretized form of these equations as used in UTCHEM can be found in Datta Gupta et $al.$ [2] and Pope et $al.$ [5]. The formulation of Datta Gupta et $al.$ [2] results in a matrix with the advantageous properties of symmetry, diagonal dominance and irreducibility. Symmetry in conjunction with irreducibility and at least weak diagonal dominance with non-negative diagonal entries ensures that the matrix is positive definite. This is an important property that allows the use of efficient algorithms for solution of the positive definite linear system of equations resulting from discretization of the pressure equation.

18.3   Vectorization of UTCHEM

The procedure used to convert UTCHEM from a scalar code to a vector code will now be discussed. Although specific to UTCHEM, part of this process will carry over to other reservoir simulators. Our objective was to fully utilize the vector-processing capability of the CRAY X-MP to enhance the speed of UTCHEM as much as possible.

Any vectorization attempt must start with examining the performance of the original code. The original version of UTCHEM is referred to as the scalar version. A list of UTCHEM subroutines is given in Table 18.1.

Cray Research Inc. provides several tools for performance evaluation purposes. SPY, FLOPTRACE and FLOWTRACE are three tools that give detailed information about performance of subroutines, functions and DO loops in FORTRAN code. SPY provides more information than the other two trace utilities, but because of its higher overhead it was used only once, at the beginning of the vectorization process. FLOPTRACE provides the speed of each routine in millions of floating point operations per second (MFLOPS). FLOWTRACE provides the time consumed by each routine and was used many times to monitor the progress of the vectorization process.

A simulation of surfactant flooding in a quarter five-spot well pattern with an area of 5.74 acres was used as a base case to evaluate the progress of the vectorization coding. The grid size was 11 × 11 × 2 (242 grid blocks). Each simulation was run for 90 constant time steps (a complete run would require 2700 time steps). Table 18.2 gives the computational results from both FLOWTRACE and FLOPTRACE for scalar UTCHEM using the CRAY X-MP/24 at The University of Texas System Center for High Performance Computing. The first column in the table presents the names of routines in alphabetical order. The second column indicates the amount of time spent in each routine, with percentage of total time given in parentheses. The third column presents the number of times each routine was called, and the last column shows the MFLOPS rate for the routines.

Theoretically, the CRAY X-MP with 9.5 ns clock period can perform 210 MFLOPS. The speed of the base case using the original scalar code was only 7.5 MFLOPS. This code was also run by turning the vectorization option off. The CPU time was 26.1 seconds compared with 23.9 seconds (Table 18.2) with vectorization turned on. Only a 9% reduction in time occurred. Thus, it was clear that changes in the code had to be made to benefit from the vector-processing capability of the CRAY X-MP.

More than 63% of the CPU time was consumed by the main program (MAIN), with most of this time spent solving the mass conservation equations. This part of MAIN was transferred to a new subroutine called CONEQ. The most important obstacle in vectorizing this segment of code was the treatment of boundaries. The solution to this problem was to initially ignore the boundaries so that a vector calculation could be made and then to compute the correct boundary points later in new DO loops. This approach was used not only for solving the conservation equations but also for other parts of the case, such as calculation of transmissibilities that required special treatment of the boundaries.

Table 18.1: Subroutines Used in Scalar UTCHEM

- MAIN   : MAIN PROGRAM. DRIVES THE PROGRAM.
             ALSO SOLVES CONSERVATION EQUATION.

Pressure Equation Subroutines

- DENSTY : PHASE DENSITY CALCULATION
- SOLMAT : SETS UP MATRIX BANDS AND RIGHT-HAND
           VECTOR
- TRAN1   : CONSTANT PORTION OF PHASE
           TRANSMISSIBILITIES
- TRANS   : PHASE TRANSMISSIBILITIES

System Solver Subroutine

- DIRECT   : LOADS THE MATRIX AND CALLS SOLVER
            ROUTINES
- LEQ1PB : (IMSL ROUTINE) SOLVER FOR POSITIVE
            DEFINITE BAND MATRIX
- LUDAPB : (IMSL ROUTINE) DECOMPOSITION OF THE MATRIX
- LUELPB : (IMSL ROUTINE) ELIMINATION

Physical Property Subroutines

- ADSORB : CHEMICAL ADSORPTION
- ALCPTN : ALCOHOL PARTITIONING COEFFICIENTS
- CSECAL : EFFECTIVE SALINITY CALCULATION
- CUBIC   : CUBIC EQUATION SOLVER
- GAMMA   : INTERFACIAL TENSION CALCULATION
- IONCNG : ION EXCHANGE CALCULATION
- PHCOMP : FLASH CALCULATIONS
- REVISE : CONVERSION, PSEUDO-TERNARY TO
            QUATERNARY
- TIELIN : TIE-LINE ITERATION
- TRAP   : PHASE TRAPPING AND RELATIVE PERMEABILITY
- TRY   : TIE-LINE ITERATION
- TWOALC : TWO ALCOHOL SYSTEM PARAMETERS
            CALCULATION
- VISCOS : PHASE VISCOSITY CALCULATION

Well Treatment Subroutines

- QRATE   : WELL FLOW RATES
- WELL   : WELL BOUNDARY CONDITIONS

I/O Subroutines

- PRINTS : PRINTS TWO-DIMENSIONAL ARRAYS
- RSTART : RESTART OPTION
- SUMTAB : SUMMARY TABLE

Table 18.2: Scalar UTCHEM for the Base Case on CRAY X-MP/24*

ROUTINE	TIME (sec)	EXECUTING (%)	CALLED	MFLOPS
ADSORB	0.052	(0.22%)	21538	2.66
ALCPTN	0.046	(0.19%)	21627	3.56
CSECAL	0.202	(0.84%)	21627	3.85
CUBIC	0.041	(0.17%)	2816	10.32
DENSTY	0.076	(0.32%)	90	26.43
DIRECT	0.023	(0.10%)	90	0.05
GAMMA	0.020	(0.08%)	2179	9.20
IONCNG	0.009	(0.04%)	21898	1.27
LEQ1PB	>	(0.00%)	90	0.00
LUDAPB	2.300	(9.64%)	90	17.63
LUELPB	0.219	(0.92%)	90	20.96
MAIN	15.039	(63.01%)	1	5.18
PHCOMP	0.365	(1.53%)	21898	7.46
PRINTS	0.116	(0.49%)	24	0.48
QRATE	0.007	(0.03%)	90	7.62
REVISE	0.006	(0.02%)	2179	10.69
RSTART	0.179	(0.75%)	1	0.67
SOLMAT	1.227	(5.14%)	90	7.23
SUMTAB	0.006	(0.03%)	1	2.42
TRAN1	>	(0.00%)	1	12.08
TRANS	1.328	(5.57%)	90	11.61
TRAP	0.985	(4.13%)	21538	9.25
VISCOS	1.610	(6.75%)	21538	10.54
WELL	0.010	(0.04%)	90	8.99
TOTAL	**23.867**		**159676** Total	**7.50** Ave.

* This CRAY X-MP/24 has a clock cycle time of 9.5 nanoseconds.
The bracket (>) indicates that the time spent is small.

The solver routines (LUDAPB and LUELPB) took more than 10% of the CPU time. As will be shown later, this percentage increases as the size of the problem increases. The solvers we used are discussed below. The subroutines WELL and QRATE took less than 0.1% of the CPU time, so they were not changed.

There were two reasons that restructuring of the code was needed to take maximum advantage of vectorization. First, many of the inner loops contained subroutine calls which prevented vectorization by the compiler. The second problem was the short vector length of the inner loops. The CRAY X-MP performs best with long inner DO loops. In the scalar version of UTCHEM, the arrays were dimensioned for each of the three dimensions NX, NY and NZ. This means that the length of the innermost loop, for example, NX, is not as long as it could be. The solution to this problem was to dimension the arrays to $NX \times NY \times NZ$, which gives the maximum vector length.

In UTCHEM, many physical properties are calculated in subroutines (ADSORB, CSECAL, VISCOS, TRAP, etc.). In the scalar code, these subroutines were called inside DO loops performing calculations grid block by grid block. This structure did not allow any of these loops to be vectorized. The solution to this problem was to push the block DO loops inside these subroutines. This not only permitted vectorization, which saved time by reducing overhead, but also substantially reduced the number of subroutine calls.

One of the most time-consuming physical property routines in UTCHEM is the phase behavior routine (PHCOMP). Several types of phase behavior can coexist in the simulated reservoir at any given time, and the situation depends on the physical problem, so it cannot be predicted in a general way. This results in loops with variable vector strides not only in PHCOMP but also in other routines which are affected by the phase behavior, such as TRAP. The approach that allowed these routines to be vectorized was the use of "gather" and "scatter" operations. The gather instruction creates arrays with the same properties for the purpose of vectorization. The scatter instruction returns the calculated values back to the original locations. In the CRAY X-MP, these operations are vectorizable using either FORTRAN or calls to subroutines GATHER and SCATTER. Cray computers are now available with hardware gather and scatter that significantly speed up these operations.

Other methods used to take as much advantage of vectorization as possible included chaining arithmetic operations and unrolling short inner loops. Using these methods, the routine DENSTY increased in speed from 26.4 MFLOPS to 157.2 MFLOPS after vectorization.

18.3.1 Vectorization Results

Table 18.3 presents a list of subroutines in vectorized UTCHEM. Table 18.4 shows the performance of each of these subroutines for the base case run using the vectorized code and the speed-up factor compared to the scalar code. The average speed was 7.5 MFLOPS (see Table 18.2) for this base case using the scalar code and 75.5 MFLOPS for the vectorized code.

Table 18.5 summarizes our computational results for five mesh sizes. The time for the scalar code increased from 23.8 seconds to 1832 seconds as the mesh was refined from $11 \times 11 \times 2$ to $44 \times 44 \times 4$. The time for the vectorized code using the same solver (Cholesky) increased from 2.51 to 241 seconds for the same refinement. The time for the vectorized code using another solver, the Jacobi Conjugate Gradient (JCG) solver, increased from 2.28 to 73.45 seconds (0.000105 sec./time-step/grid block), again for this same refinement. This is a speed-up factor of 25 compared to the optimized scalar code. A complete chemical flood (1500 days in this case) using this $44 \times 44 \times 4$ mesh and the scalar code would require about 15.3 hours on the CRAY X-MP compared to only 0.61 hours using the vectorized code. The savings is even greater for more complex simulations such as those involving more complicated phase behavior.

18.3.2 Solvers

The implicit solution of the pressure equation in UTCHEM gives rise to a system of linear equations of the form $A\boldsymbol{x} = \boldsymbol{b}$, where A is a symmetric positive definite banded matrix. Numerical solution of this system of equations is very time-consuming, especially as the number of equations becomes large. Mamun *et al.* [3] reported a comparison of iterative methods on vector processors which showed that the JCG method was the fastest of the methods tested on several standard reservoir engineering simulators. We also have found that the JCG method is faster than the one we previously used in UTCHEM, which was the Cholesky method (see Table 18.6). The improvement increases as the size of the problem increases. Another advantage of the JCG method over the Cholesky method is that it requires less storage.

18.4 Microtasking Results

As part of our continuing effort to improve the speed of UTCHEM, we tested the use of Microtasking in this code. Microtasking reduces the wall clock time rather than the CPU time by using more than one processor at the same time. This is an advantage when very long runs are needed. Microtasking is possible with very little change in the original code and, in our

Table **18.3**: Subroutines Used in Vectorized UTCHEM

* MAIN	: MAIN PROGRAM. DRIVES THE PROGRAM.
* CONEQ	: SOLVES THE CONSERVATION EQUATIONS

Pressure Equation Subroutines

* DENSTY	: PHASE DENSITY CALCULATION
* SOLMAT	: SETS UP MATRIX BANDS AND RIGHT-HAND VECTOR
* TRAN1	: CONSTANT PORTION OF PHASE TRANSMISSIBILITIES
* TRANS	: PHASE TRANSMISSIBILITIES

System Solver Subroutine

* JCG	: JACOBI CONJUGATE GRADIENT SOLVER

Physical Property Subroutines

* ADSORB	: CHEMICAL ADSORPTION
* ALCPTN	: ALCOHOL PARTITIONING COEFFICIENTS
* CSECAL	: EFFECTIVE SALINITY CALCULATION
* CUBIC	: CUBIC EQUATION SOLVER
$ IONCNG	: ION EXCHANGE CALCULATION
* PHCOMP	: FLASH CALCULATIONS
%* REVISE	: CONVERSION, PSEUDO-TERNARY TO QUATERNARY
% SGAMMA	: INTERFACIAL TENSION CALCULATION (SCALAR)
$ SREVISE	: SCALAR REVISE
$ TIELIN	: TIE-LINE ITERATION
* TRAP	: PHASE TRAPPING AND RELATIVE PERMEABILITY
$ TRY	: TIE-LINE ITERATION
$ TWOALC	: TWO ALCOHOL SYSTEM PARAMETERS CALCULATION
%*SINGLE	: SINGLE PHASE REGION CALCULATION
%*VGAMMA	: INTERFACIAL TENSION CALCULATION
* VISCOS	: PHASE VISCOSITY CALCULATION

Well Treatment Subroutines

$ QRATE	: WELL FLOW RATES
$ WELL	: WELL BOUNDARY CONDITIONS

I/O Subroutines

$ PRINTS	: PRINTS TWO-DIMENSIONAL ARRAYS
$ RSTART	: RESTART OPTION
$ SUMTAB	: SUMMARY TABLE

*	: Vectorized
%	: New Subroutine
$	: Revised and Optimized

Table 18.4: Vectorized UTCHEM with JCG for Base Case
on CRAY X-MP/24*.

ROUTINE	TIME (sec)	EXECUTING (%)	CALLED	MFLOPS	SPEED-UP
ADSORB	0.015	(0.66%)	89	70.79	3.47
ALCPTN	0.013	(0.55%)	178	42.41	3.54
CONEQ	0.873	(38.18%)	89	102.76	—
CSECAL	0.021	(0.93%)	178	66.14	9.62
CUBIC	0.007	(0.29%)	89	95.39	5.86
DENSTY	0.013	(0.55%)	90	157.23	5.85
IONCNG	0.001	(0.05%)	263	20.90	9.00
JCG	0.303	(13.25%)	90	108.49	8.39
MAIN	0.074	(3.24%)	1	16.15	15.88
PHCOMP	0.083	(3.65%)	263	48.52	4.35
PRINTS	0.116	(5.06%)	24	0.47	1.00
QRATE	0.007	(0.31%)	90	7.78	1.00
REVISE	0.003	(0.15%)	171	31.27	2.00
RSTART	0.179	(7.38%)	1	0.67	1.00
SINGLE	0.001	(0.06%)	83	1.92	—
SOLMAT	0.084	(3.67%)	90	75.52	14.61
SUMTAB	0.006	(0.26%)	1	0.83	1.00
TRAN1	>	(0.01%)	1	58.85	—
TRANS	0.107	(4.67%)	90	47.83	12.41
TRAP	0.126	(5.49%)	89	78.54	7.82
VGAMMA	0.004	(0.15%)	171	41.23	5.00
VISCOS	0.241	(10.53%)	89	73.28	6.68
WELL	0.010	(0.45%)	90	8.94	1.00
TOTAL	**2.287**		2320 Total	75.51 Ave.	10.44 Overall

* This CRAY X-MP/24 has a clock cycle time of 9.5 nanoseconds.
The bracket (>) indicates that the time spent is small.

Table 18.5: CPU Times for UTCHEM on CRAY X-MP/24*.

GRID SIZE NX×NY×NZ	SC (sec)	VCH (sec)	VJCG (sec)	SPEED-UP SC/VCH	SPEED-UP SC/VJCG
11 × 11 × 2	23.86	2.51	2.28	9.50	10.43
11 × 11 × 4	55.31	5.30	4.62	10.43	11.97
22 × 22 × 4	274.73	27.71	17.12	9.91	16.04
33 × 33 × 4	736.83	93.09	39.60	7.91	18.60
44 × 44 × 4	1832.09	241.08	73.45	7.60	24.94

NX : No. of Grid Blocks in X-Direction
NY : No. of Grid Blocks in Y-Direction
NZ : No. of Grid Blocks in Z-Direction
SC : Scalar Code with Cholesky Solver
VCH : Vectorized Code with Cholesky Solver
VJCG : Vectorized Code with Jacobi Conjugate Gradient Solver
* This CRAY X-MP/24 has a clock cycle time of 9.5 nanoseconds.

experience, has little effect on the total computational time if the code is run on a Cray computer using only one processor.

The most time-consuming parts of UTCHEM were identified using FLOW-TRACE on the Cray X-MP. Table 18.4 shows these results for our base-case simulation with a 11 × 11 × 2 grid. Only those subroutines which took more than 5 percent of the total time were Microtasked. The details of how these subroutines were Microtasked can be found in Pashapour et al. [4].

Table 18.7 shows our results for up to four processors on a CRAY X-MP/48 at Mendota Heights under dedicated conditions. As seen from this table, the speed-up factors varied from 1.33 to 1.55 for two processors, 1.42 to 1.85 for three processors, and 1.49 to 2.06 for four. The speed-up factor was calculated as the ratio of the wall clock time for a uni-processor system to the wall clock time for a multiple-processor system. Higher speed-up factors were achieved by increasing the number of grid blocks. This is mainly due to the increase in vector length of inner loops within most of the control structures executed in parallel. Examination of the individual Microtasked routines revealed that speed-up factors of up to 3.8 were obtained using four processors. However, the overall speed-up achieved using four processors was only as high as 2.06. This is attributed to the following: 1) load balancing, 2) task granularity, 3) memory contention, and 4) balance between Microtasking and vectorization. Table 15.8 presents the wall clock time and speed-up factor for the Microtasked JCG solver using one to three processors on a CRAY X-

Table 18.6: Comparison of Direct and Iterative Solvers for UTCHEM Simulator with 90 Constant Time Steps on CRAY X-MP/24

Grid Size	Time (sec)				Percentage of total computation time consumed by the solver				MFLOPS		
	Cholesky				Cholesky				Cholesky		
	BNDCHL*	BNDCHLS*	Total	JCG	BNDCHL*	BNDCHLS*	Total	JCG	BNDCHL*	BNDCHLS*	JCG
11 × 11 × 2	0.36	0.07	0.44	0.33	15.78	3.30	19.08	14.85	32	24	108
11 × 11 × 4	1.36	0.18	1.54	0.99	25.72	3.38	29.10	21.42	62	41	120
22 × 22 × 4	13.02	1.09	14.11	4.22	47.03	3.93	50.96	24.64	104	55	141
33 × 33 × 4	59.57	3.31	62.88	11.43	64.00	3.55	67.55	28.87	115	61	143
44 × 44 × 4	-	-	-	24.04	-	-	-	32.74	-	-	146

- Not enough memory to run this case using FLOWTRACE or FLOPTRACE.
* BNDCHL and BNDCHLS are the vectorized solvers.

Table 18.7: Wall Clock Time (sec) and Speed-up for Microtasked UTCHEM on CRAY X-MP/48*

Grid Size	1-Proc.	2-Proc./SP	3-Proc./SP	4-Proc./SP
11 × 11 × 2	2.02	1.52/1.33	1.42/1.42	1.36/1.49
11 × 11 × 4	4.28	3.05/1.40	2.69/1.59	2.55/1.68
22 × 22 × 4	15.38	10.32/1.49	8.80/1.75	7.95/1.94
33 × 33 × 4	35.49	23.13/1.53	19.49/1.82	17.60/2.02
44 × 44 × 4	66.13	42.53/1.55	35.79/1.85	32.17/2.06

SP: speed-up factor
* This CRAY X-MP/48 has a clock cycle time of 8.5 nanoseconds.

Table **18.8**: Wall Clock Time (sec) and Speed-up for Microtasked JCG Solver on CRAY X-MP/24* for UTCHEM Simulator

GRID SIZE	1-PROC.	2-PROC./SP	3-PROC./SP
$11 \times 11 \times 2$	0.39	0.30/1.30	0.32/1.21
$11 \times 11 \times 4$	1.09	0.78/1.39	0.72/1.51
$22 \times 22 \times 4$	4.30	2.65/1.62	2.32/1.85
$33 \times 33 \times 4$	11.45	6.55/1.76	5.64/2.03
$44 \times 44 \times 4$	24.47	13.80/1.77	11.74/2.08

SP :   Speed-up factor

* This CRAY X-MP/48 has a clock cycle time of 8.5 nanoseconds.

MP/48. The computational results using the fourth processor did not show any further improvement, possibly due to memory contentions or competition between vectorization and parallel processing. Work is underway to obtain a greater speed-up factor for the UTCHEM simulator using parallel processing (Pashapour *et al.* [4]).

18.5   Conclusions

Speed-up factors of up to 25 demonstrated that a combination of vectorization and the use of a better solver resulted in a significant savings in computational time for our chemical flooding simulator, UTCHEM. To achieve this, the structure of the program had to be changed. The most significant savings resulted from vectorization of the solution of the concentration equations, the physical property routines, and the solver. Vectorization of UTCHEM eliminated most of the data dependencies in the code and helped in recognizing the sections that were totally independent of each other and could be executed simultaneously. Ease in implementation of Microtasking made it possible to exploit parallel processing on the CRAY X-MP. Field scale simulations are not unreasonably expensive now. Numerous examples of these simulations can be found in Allison *et al.* [1] and Saad *et al.* [7].

Nomenclature

c_f	= pore compressibility
$\bar{c}_i$	= adsorbed concentration of component i in units of volume per pore volume

c_{ij}	$=$ concentration of component i in phase j
c_i^0	$=$ compressibility of component i
K_{ij}	$=$ dispersion coefficient of component i in phase j
n_c	$=$ number of components
n_p	$=$ number of phases
P	$=$ pressure
P_R	$=$ reference pressure
Q_i	$=$ source/sink term
S_j	$=$ saturation of phase j
t	$=$ time
u_{xj}, u_{yj}, u_{zj}	$=$ Darcy velocity of phase j in x-, y-, and z-directions, respectively
ϕ	$=$ porosity

Acknowledgments

Several helpful discussions with Jeff Nicholson of Cray Research Inc., Mendota Heights, are gratefully acknowledged. This study was made possible by financial support from Cray Research Inc., the U.S. Department of Energy, and the participating companies at the Center for Enhanced Oil and Gas Recovery Research at The University of Texas at Austin: Amoco, Arco, ADREF, Chevron, Conoco, Cray Research Inc., Digital Equipment Corp., Elf Aquitaine Production, Enterprise Oil, Institute for Energy Technology, IN-TEVEP, Japan National Oil Corp., Japan Petroleum Exploration Co., Mobil, Norsk Hydro, Oxy U.S.A., Phillips Petroleum, Rogaland Research Institute, Shell, Standard Oil, Statoil, Sun E & P, Tenneco, Texaco and Unocal. Computing resources for this work were provided by The University of Texas System Center for High Performance Computing and Cray Research Inc., Mendota Heights.

References

[1] Allison, S. B., G. A. Pope, and K. Sepehrnoori, "Analysis of Field Tracers for Reservoir Description," *Journal of Petroleum Science and Engineering*, to appear, 1989.

[2] Datta Gupta, A., G. A. Pope, K. Sepehrnoori, and R. L. Thrasher, "A Symmetric, Positive Definite Formulation of a Three-Dimensional Micellar/Polymer Simulator," *SPE Reservoir Engineering*, 1, 6, 622-632, 1986.

[3] Mamun, C. K., G. A. Pope, and K. Sepehrnoori, *A Comparison of Iterative Methods for Reservoir Simulation Problems on Vector Processors,* Report No. 86-2, Center for Enhanced Oil and Gas Recovery Research, The University of Texas at Austin, 1986.

[4] Pashapour, A., G. A. Pope, K. Sepehrnoori, and G. Shiles, "Application of Parallel Processors in Compositional Simulation," (in preparation) (1989).

[5] Pope, G. A., L. W. Lake, and K. Sepehrnoori, *Modelling and Scale-up of Chemical Flooding,* U.S. Department of Energy DOE/BC/10846-6, 1987.

[6] Saad, N., G. A. Pope, and K. Sepehrnoori, "Application of Higher Order Methods in Compositional Simulation," submitted to *SPE Reservoir Engineering,* June, 1988.

[7] Saad, N., G. A. Pope, and K. Sepehrnoori, "Simulation of Big Muddy Surfactant Pilot," *SPE Reservoir Engineering,* 4, 1, 24–34, 1989.

Index